Human Diseases Caused by Viruses

HUMAN DISEASES CAUSED BY VIRUSES

RECENT DEVELOPMENTS

EDITORS

Henry Rothschild
Fred Allison, Jr.
Calderon Howe

COORDINATING EDITOR

Charles F. Chapman

New York
OXFORD UNIVERSITY PRESS
1978

Library of Congress Cataloging in Publication Data

Main entry under title:

Human diseases caused by viruses.

   Bibliography: p.
   Includes index.
   1.   Virus diseases.   I.   Rothschild, Henry, 1932-
II.   Allison, Fred, 1922-   III.   Howe, Calderon,
1916-
RC114.5.H86   616.9′2   76-57482
ISBN 0-19-502286-6

Printed in the United States of America

DEDICATION

The late G. John Buddingh, during his long career as head of the Department of Microbiology, Louisiana State University School of Medicine, established a tradition in New Orleans for research on viruses and the infections they cause.

Trained first in medicine and then in pathology under the aegis of Ernest W. Goodpasture, Dr. Buddingh had been one of the first to adapt the embryonated chicken egg to the cultivation of viruses. This development represented an important breakthrough in methodology for the primary isolation and quantitation of viral agents and the study of host/parasite interactions.

His scholarly pursuits in many areas of biomedical science kindled in his students the type of intellectual curiosity that was his own hallmark. For his contributions to virology, we affectionately dedicate this book to John Buddingh.

CONTRIBUTORS

FRED ALLISON, JR., MD   The Edgar Hull Alumni Professor of Medicine and Microbiology; Head, Department of Medicine, Louisiana State University Medical Center, New Orleans, La.

PAUL H. BLACK, MD   Associate Physician, Massachusetts General Hospital; Associate Professor, Department of Medicine, Harvard Medical School, Boston, Mass.

CHARLES F. CHAPMAN, PhB   Director, Editorial Office, Louisiana State University Medical Center, New Orleans, La.

D. CARLETON GAJDUSEK, MD   Chief, Central Nervous System Studies Laboratory, National Institute of Neurological Diseases and Stroke, National Institutes of Health, Bethesda, Md.

DOROTHY M. HORSTMANN, MD   John Rodman Paul Professor, Departments of Epidemiology, Public Health, and Pediatrics, Yale University School of Medicine, New Haven, Conn.

CALDERON HOWE, MD   Professor and Head, Department of Microbiology, Louisiana State University Medical Center, New Orleans, La.

GEORGE GEE JACKSON, MD   Professor, Department of Medicine; Head, Section of Infectious Diseases, Abraham Lincoln School of Medicine, University of Illinois, Chicago, Ill.

CONTRIBUTORS

GEORGE KHOURY, MD Head, Virus Tumor Biology Section, Laboratory of DNA Tumor Viruses, National Cancer Institute, National Institutes of Health, Bethesda, Md.

EDWIN D. KILBOURNE, MD Professor and Chairman, Department of Microbiology, Mount Sinai School of Medicine, New York, N.Y.

SAUL KRUGMAN, MD Professor, Department of Pediatrics, New York University School of Medicine, New York, N.Y.

JAMES P. LUBY, MD Associate Professor, Department of Medicine; Chief, Infectious Disease Division, The University of Texas Health Science Center, Dallas, Tex.

JOSEPH L. MELNICK, PhD Distinguished Service Professor and Chairman, Department of Virology and Epidemiology, Baylor College of Medicine, Houston, Tex.

I. GEORGE MILLER, MD Professor, Department of Pediatrics and Epidemiology, Yale University School of Medicine, New Haven, Conn.

JAMES A. ROBB, MD Associate Professor, Department of Pathology, University of California, San Diego, School of Medicine, La Jolla, Calif.

LEON ROSEN, MD PhPH, DrPH Director, Pacific Research Section, Laboratory of Parasitic Diseases, National Institute of Allergy and Infectious Diseases, National Institutes of Health, Honolulu, Hawaii

HENRY ROTHSCHILD, MD, PhD Professor, Departments of Medicine and Anatomy, Louisiana State University Medical Center, New Orleans, La.

CHARLES V. SANDERS, Jr., MD Associate Professor, Departments of Medicine and Microbiology; Chief, Infectious Diseases Section, Louisiana State University Medical Center, New Orleans, La.

CONTRIBUTORS

GILBERT M. SCHIFF, MD   Professor, Department of Medicine; Director, Infectious Disease Division, University of Cincinnati Medical Center, and Director, Christ Hospital Institute of Medical Research, Cincinnati, Ohio

JERRY W. SMITH, PhD   Associate Professor, Department of Microbiology, Louisiana State University Medical Center, New Orleans, La.

FOREWORD

In 1892, Ivanovski observed that a "filterable agent" caused mosaic disease in tobacco plants. This was the otherwise unheralded birth of the science of virology, which has grown exponentially, drawing its nutriment first from early endeavors of Löffler and Frosch, Reed, Rous, Twort, and D'Herelle, later from Woodruff and Goodpasture, Stanley, Bittner, and others, and in the contemporary period, from such investigators as Gajdusek who have redefined the orginal concept of a virus.

With expanding research have come increased information and improved methods for diagnosis and treatment of viral diseases. However, the pace of progress in the field is so swift that the message published today on yesterday's observation is in danger of being obsolete tomorrow. Because of that danger, great care and concern were given to the selection of topics and speakers for the conference from which this volume derives.* The objective was to present basic and current information for clinicians—particularly family physicians, pediatricians, and internists—that would have applicability in practice.†

Because the scope of virology is broad and many unresolved and complex problems remain, our intent was to invite experts to contribute information on subjects that are practical and pertinent. Some diseases, such as smallpox, have been virtually eliminated from the Western world and, hence, were not included in our considerations. They have been superseded by subjects of greater currency and interest

---

* "Human Disease Caused by Viruses: Recent Developments," Louisiana State University School of Medicine, 12 and 13 November, 1976.
† For a more comprehensive review of viruses, we recommend the standard textbooks, such as Jawetz, Melnick, and Adelberg's *Review of Medical Microbiology,* Los Altos, Calif., Lange Medical Publishers, 1976; or the second edition of Fenner and White's *Medical Virology,* New York, Academic Press, 1976.

or by those in which recent developments are more definitive and reportable.

The reader is warned that we have included some topics that, at present, have limited clinical application. They are included to provide a background on recently emerging aspects of virology. To understand animal virology one must understand molecular virology, the latter having its roots in the former. The clinician may find this rather specialized information somewhat difficult to read. We have therefore supplied a glossary at the end of this volume to supplement the text. We hope, nonetheless, that the technical and theoretical information will be helpful for understanding recent developments and the mechanisms underlying diseases caused by viruses. Our intent has been to provide reference material and to illuminate significant findings, changes, and trends in basic and applied virology.

We have used a light hand in editing each chapter, our guiding principle having been to preserve the identity of each contributor with his work while shaping the whole to achieve a consistent and readable text. References were carefully selected to meet our stated objective, some chapters having many references and others relatively few. We have included portions of the pertinent discussions that followed the presentation of each subject at the conference. Our efforts will have been well served if the practitioner, teacher, and student find the information useful.

We thank Drs. Harry E. Dascomb and Charles V. Sanders for assisting in editing the book; Dr. Silas E. O'Quinn for his encouragement and support; and Drs. Barnett L. Cline, Carl J. Dicharry, Philip Dolan, Nicholas Gagliano, William R. Gallaher, Brown C. Mason, Harold Trapido, and William L. Williams for their expertise and helpful review of manuscripts.

October 1977

Fred Allison, Jr.
Charles Chapman
Henry Rothschild

Contents

1

The Evolution of Virology

CALDERON HOWE

To gather some historical perspective on the relation of viruses to human disease, I draw your attention to a remarkable treatise published in 1935 that represented all the information then available about infectious diseases and their causation. I refer to *Agents of Disease and Host Resistance,* by Frederick P. Gay and his associates at Columbia University (1). To quote Dr. Gay: "We may conclude, then, that there are to date approximately 469 separate and specific living disease agents that are recognized morphologically with greater or less exactitude; and in addition some 46 agents which are presumably living but which have not as yet been seen." The latter small group he referred to as the "filterable viruses," including certain intermediate forms that are now recognized as being akin to bacteria rather than to true viruses, for example, the agents of psittacosis and trachoma (*Chlamydia*).

THE BEGINNING

In the accompanying diagram (Fig. 1.1), "the abscissa marks the lapse of time in the history of discovery and the ordinates the increments of advance in the term of number of new organisms discovered in any particular year," right up to the point at which Dr. Gay's book was published. The steep acceleration for the period 1875 to 1900 represents the "golden age of bacteriology." Viruses as we know them today are conspicuous by their absence. However, a few signal observations made around the turn of the century have been added to the diagram to indicate that the "golden age of virology" was soon to begin. At the very end of the 19th century, Löffler and Frosch (2) described the nonbacterial cause of foot-and-mouth disease in ungulates. At almost the same time, Beijerinck (3) described what he called a "contagium virum fluidum," a euphemism for a filterable nonbacterial agent that he demonstrated was the cause of tobacco mosaic disease. The virus was crystallized in 1935 by Stanley (4). The latter landmark studies represented the first chemical definition of a virus.

In 1908, Karl Landsteiner, who was also the father of modern immunochemistry, was the first to transmit poliomyelitis by intracerebral inoculation of bacteria-free brain tissue in monkeys (5). Although filterable viruses were increasingly recognized as nonbacterial causes of well-known disease syndromes, early attempts at propagation and passage were of necessity limited to laboratory animals. However, in the late 1920s and 1930s, Ernest Goodpasture began to exploit the embryonated chicken egg as a medium for the propagation of viruses.

3

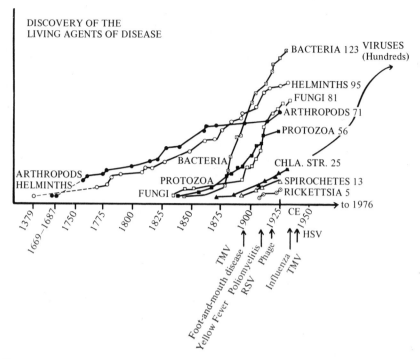

Figure 1.1. Cumulative discovery of major categories of infectious disease agents. Chla. Str. = chlamydozoa-strongyloplasm, obsolete terms applied to "inclusion bodies," the nature of which was unknown but which were recognized by light microscopy as being pathognomonic of certain viral infections. Through electron microscopy, "inclusion bodies" have now been shown to be aggregates of viral particles or subunits that, in association with morbid changes in the host cell, constitute the cytopathic effects characteristic of each major viral group. TMV = tobacco mosaic virus; RSV = Rous sarcoma virus; CE = chicken embryo; HSV = herpes simplex virus. (Modified from ref. 1.)

In those early endeavors, Dr. Goodpasture had the enthusiastic collaboration of our colleague and friend, the late G. John Buddingh, whose own contributions to this field were so important in the development of animal virology (6). Goodpasture and his associates in America, and Burnet in Australia, laid the foundation for the enormous amount of work that followed during the next two decades on the biology and epidemiology of the influenza viruses, herpes simplex virus (HSV), and other agents, including the first report on the preparation

4

of smallpox vaccine in the chicken embryo. Influenza virus itself, first isolated in ferrets by Smith, Andrewes, and Laidlaw in 1931 (7), was soon adapted to the embryonated egg, the technique of which is still the basis for vaccine development and production, even in the present era of molecular politics.

During World War I, Twort and D'Herelle had independently described the phenomenon of spontaneous lysis of bacterial colonies (Fig. 1.2). In the light of our present knowledge about bacteriophage, plasmids, and related phenomena, the prominence of *Shigella* among the organisms even then recognized as having a "lytic" property is

Figure 1.2.   Drawing at low-power magnification of a diffuse plate growth (dark background) of *Escherichia coli,* illustrating bacteriolysis (clear areas) and resistant colonies within the lytic plaques. (From ref. 1.)

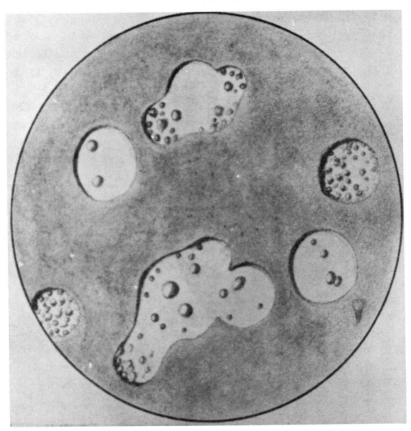

worth noting. D'Herelle gave the name "bacteriophage" to the factor that lysed bacteria and that was itself recognized as a living autonomous unit (8). The molecular biology of bacterial viruses has been the pathfinder for most concepts of modern animal virology.

Other major trends in biomedical science in the 1940s and 1950s included the development of electron microscopy, along with the technique of thin sectioning to reveal tissue structures not visible with the light microscope; the clear demonstration that nucleic acids are the "stuff that genes are made of"; the delineation of the structure of deoxyribonucleic acid (DNA) and recognition of the genetic code; and the advent of antibiotics, which, because of their low toxicity for animal cells and potent inhibitory action against common contaminating bacteria, were fundamental to the emergence of cell and tissue culture techniques. All these developments added great impetus to the evolution of modern virology.

With the foregoing, one must keep in mind the extraordinary and concurrent advances that have occurred in immunobiology, all of which have bearing on understanding the pathogenesis of viral diseases and their prevention. From the beginning of immunochemistry in the 1930s with Karl Landsteiner, Michael Heidelburger, and others, immunology has now reached the point at which most immune mechanisms, particularly specific interactions with agents of infectious disease and the manifestations of allergy and hypersensitivity, are almost entirely explicable at the cellular and molecular levels. The structure and genetics of immunoglobulins, complement, and various cellular components in the ontogeny and phylogeny of the immune response to foreign antigens constitute a body of information that continues to evolve rapidly and is intimately related to all disease-producing agents, be they viruses, bacteria, fungi, or protozoa.

THE MATURATION

Viruses are strictly dependent for survival and replication on living cells of the host (Fig. 1.3). In the present state of our knowledge of cell biology, detailed study of the molecular events and interactions that follow viral infection has been possible. We now know in some detail about viral attachment, replication, and release and how these phenomena interact with the immunological environment. Receptors

6

THE CELL IN RELATION TO

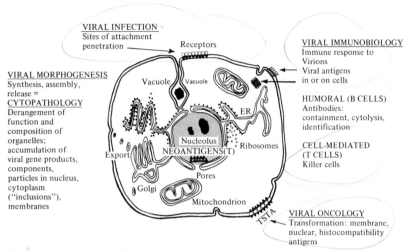

Figure 1.3. Virus-cell interactions schematized.

on cells are highly specific and therefore account, in part at least, for some of the tissue tropisms that have long been recognized. The synthesis and maturation of the viral particle vary with each system. Cytopathology as recognized today by light and electron microscopy is the totality of changes resulting from infection and the synthesis of new viruses.

Before 1950, descriptions of viral cytopathology had been limited to what was visible by light microscopy as cellular "inclusions." For these, the now obsolete terms "chlamydozoa" and "strongyloplasm" (Fig. 1.1, Chla. Str.) were coined, largely in ignorance of their true nature, save for the specific relation they held to infection with certain "filterable viruses," for example, rabies, vaccinia, and herpes simplex viruses. With the advent of cell culture and other techniques of modern virology, the molecular analysis of "inclusions" and other virus-induced cytological abnormalities became possible. In consequence, hundreds of new viruses have been recognized, many clearly causative of disease, others in the category of so-called orphan viruses. Thus has Dr. Gay's original list of 46 "unseen agents" been vastly expanded.

From close biochemical and biophysical scrutiny of newly discovered agents, a broadening base has developed on which to establish hitherto unsuspected interspecies relationships within major groups of viruses. Concomitantly, the enormous body of data on viral genetics has had direct applicability to the development of vaccines, as exemplified by the influenza problem. Discoveries have extended to new plant and insect viruses, whose relationship to human disease is not yet clear but undoubtedly will emerge as new developments occur. Even the present fundamental definition of a virus may need revision when we consider the problem of viroids.

THE PRESENT

Although much has now been learned through either light or electron microscopy about the events in the cell that are responsible for visible changes, or cytopathology, relatively little is precisely known of the mechanisms whereby viral agents upset homeostatic equilibrium in the intact host. These processes in general are poorly understood except when a particular target cell having a distinctive function is involved by viral infection, as is the case in poliomyelitis or hepatitis. How viruses otherwise produce what is vaguely called "toxemia," cause fever and other signs and symptoms of disease, or even result in death remains obscure.

As with bacteria and other agents of disease, viruses are complex mixtures of antigens. The intact virion has an outer protein coat, either a naked capsid or an envelope, that interacts with components of the immunological response evoked to it. The most obvious mechanism by which antibody works to prevent infection is by blocking early stages of attachment, internalization, or penetration of virions. Other interactions with antibodies may be destructive for the most part rather than protective. In certain infections, antigen/antibody complexes themselves may be responsible for pathological changes. The cell membrane is in contact with the immune mechanism. Hence, changes in the membrane that result from viral synthesis may bring the cell under attack by antibody and complement or by "killer" cells. The characteristics of immune responses to infection in turn provide the basis for serodiagnosis and the rationale for immunization.

The complexities of the cell-mediated arm of the immune response to viral infection are just beginning to be understood. The effect of certain viral infections on the capacity of the immune system as a whole to respond to heterologous antigens is of great interest. For example, the anergy that accompanies active measles virus infection is a well-known clinical phenomenon that is still not fully explained. The rash itself is considered to be a manifestation of localized T-cell activity. Diseases in which autoimmune phenomena occur include hepatitis, systemic lupus erythematosus and its experimental counterpart in New Zealand mice, and subacute sclerosing panencephalitis associated with antecedent measles infection and now also recognized as possibly being a late sequel to the rubella syndrome.

In the expanding realm of viral oncology, the intimate relationship between viral replication and the immune system is reflected in the complex processes of cell transformation by oncogenic viruses that are integrated into the cellular genome and provoke the appearance of tumor-specific neoantigens. These considerations underlie evidence for the suspected association of certain viruses with human cancer.

In this panorama, one should not forget the direct importance of bacterial viruses (bacteriophages) in human disease. For instance, viral genes, introduced into *Corynebacterium diphtheriae* by lysogenic conversion, code for the production of diphtherial toxin. A similar mechanism accounts for the production of erythrogenic toxin by group A beta-hemolytic streptococci and for the production of some staphylococcal exotoxins. Viral genes control other bacterial characteristics such as the O antigens of *Salmonella*. Other areas involving bacteriophage and bacterial genetics are also of great importance to human disease. Extrachromosomal factors (resistance-transfer factors and plasmids) that govern the expression of antibiotic resistance are becoming increasingly important because of the tons of antibiotics now being used with abandon all over the world. At the moment, we seem to be losing our capacity to cope with the consequences of the alarming selection of "Andromeda strains" of bacteria by contamination of both internal and external environments with antibiotics. Finally, the development of so-called recombinant research will soon bridge the gap between bacterial and mammalian biology.

In its present state, virology appears to be a jumble of information, some of it exact, some of it not too exact, which at once exasperates and tantalizes the investigator and the student of medicine but which also explains why it is no longer possible to present any of these topics in neat, well-defined packages. In fact, the information explosion in virology, immunology, and cell biology is of such magnitude that not a day passes but that the news media make some mention of these active frontiers of science. When neuraminidase, hemagglutinin, and the genetic code are mentioned in that redoubtable journal, the *New York Times,* one can really begin to believe in their existence! It is of particular importance that physicians on the firing line be sensible of the fact that unfettered, wide-ranging basic research has brought vast, though unpredictable dividends that will invariably undergird our means of helping sick people. In this very real sense, basic research is directly applicable to the practice of clinical medicine.

REFERENCES

1. Gay FP: *Agents of Disease and Host Resistance.* Springfield, Ill, Charles C Thomas, 1935, p 146.
2. Löffler F, Frosch P: Berichte der Kommission zur Erforschung der Maul und Klauenseuche bei dem Institut für Infektionskrankheiten in Berlin. *Zentralbl Bakteriol* I, 23:371–391, 1898.
3. Beijerinck MW: Über ein Contagium vivum fluidum als Ursache der Fleckenkrankheit der Tabaksblätter. *Zentralbl Bakteriol* II, 5:27–33, 1899.
4. Stanley WM: Isolation of a crystalline protein possessing the properties of tobacco-mosaic virus. *Science* 81:644–645, 1935.
5. Landstein K, Popper, E: Microscopische Präparate von einem menschlichen und zwei Affenrückenmarken. *Wien Klin Wochenschr* 21:1830, 1908.
6. Buddingh GJ: The chick embryo for the study of infection and immunity (Editorial). *J Infect Dis* 121:660–663, 1970.
7. Smith W, Andrewes CH, Laidlaw PP: A virus obtained from influenza patients. *Lancet* 2:66–68, 1933.
8. D'Herelle F: Sur un microbe invisible antagoniste des bacilles dysentériques. *CR Acad Sci* [D] (Paris) 165:373–375, 1917.

2

The Classification of Viruses

JOSEPH L. MELNICK

Until about 1950, little was known about viruses other than their pathogenic effect in causing diseases, and thus any efforts at classification tended to focus on host response rather than on properties of the virus particle. At present the end of an important phase of discovery and characterization of animal viruses is approaching. The knowledge gained has made it possible to establish and broadly define groupings for these agents. Apparently, most of the major groups of viruses of vertebrates—at least those affecting man and the animals important to man—have been recognized and described. Many of these virus groupings, initially established on tentative and provisional bases, now appear to form "real" families and genera, in which the members are indeed related in fundamental ways. For example, the validity of the original grouping of the enteroviruses based on an enteric habitat and small size is being borne out by current studies that utilize sophisticated techniques of modern molecular virology to compare the genetic makeup of different members of the group and their mode of replication.

Because this chapter is rather technical, it may have only limited interest for the clinician. However, it is the aim of this chapter to provide essential, current information on the classification of viruses, particularly for those seeking to clarify terminology, and to give a basis of reference for information in the following chapters.

Emphasis has shifted from sketching the broad outlines of the virus kingdom based on disease causation to filling in essential details about the viruses themselves. This shift is reflected by the change in the name of the International Committee on Nomenclature of Viruses (ICNV) to the International Committee on Taxonomy of Viruses (ICTV). The ICNV was established in 1966 at a historic meeting in Moscow, the very city where viruses were first discovered three quarters of a century earlier by Ivanovski. The first report of the ICNV was published in 1971 (1). Work of the study groups and subcommittees of the ICTV is proceeding, and they are reporting an increasing number of taxonomic articles in their special areas of virology. Reports of these groups appear regularly in *Intervirology,* the journal of the Virology Section of the International Association of Microbiological Societies. Official decisions of the ICTV made at its meetings held during the Third International Congress for Virology have recently been summarized (2), and a full report of the results of the last five years of work by the ICTV and its committees has been published as a special double issue of *Intervirology* (3).

Figures 2.1 and 2.2 serve as useful reference points in the following discussion of classification based on properties of the virus particles,

13

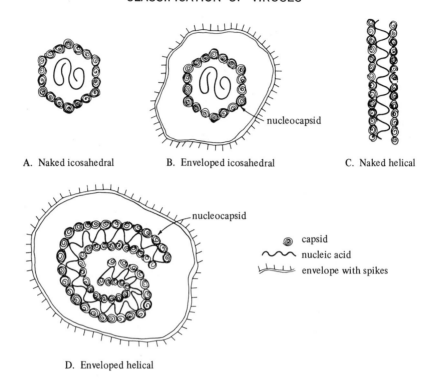

A. Naked icosahedral   B. Enveloped icosahedral   C. Naked helical

D. Enveloped helical

Figure 2.1. Schematic diagram of simple forms of virions and their components. The naked icosahedral virions resemble small crystals; the naked helical virions resemble rods with a fine regular helical pattern in their surface. The enveloped icosahedral virions are made up of icosahedral nucleocapsids surrounded by the envelope; the enveloped helical virions are helical nucleocapsids bent to form a coarse, often irregular coil within the envelope. (From ref. 4.)

and comparison of these two figures also gives some indication of the rapid accumulation of knowledge about virus composition and structure. Figure 2.1 is taken from a text (4) published about 10 years ago; it remains fundamentally applicable in current virology. However, Figure 2.2, a diagram prepared 10 years later (5), not only includes additional information that has been gained about viruses but also illustrates the wide variety of size and structure that is found among the viruses of vertebrates.

In addition to the ICTV and its study groups, general resources for the data presented here include my own reviews on taxonomy of ani-

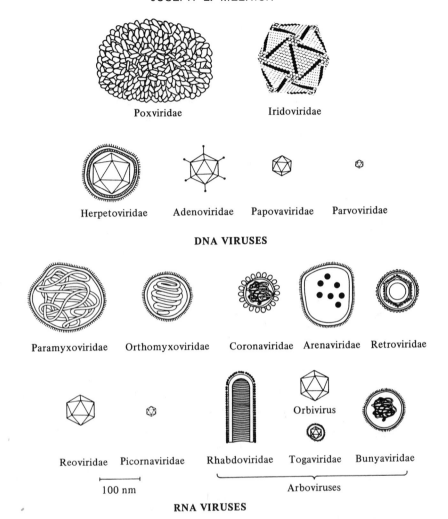

Figure 2.2. Diagram illustrating the shapes and relative sizes of animal viruses of the major families (bar = 100 nm). (From ref. 5.)

mal viruses, published regularly in *Progress in Medical Virology* since 1966 (6), as well as a current review by the chairman of the Vertebrate Virus Subcommittee of the ICTV (7). Figures 2.3 through 2.7 are schematic diagrams showing separation of viruses of vertebrates into 16 families (6). Figure 2.3 describes viruses having a DNA genome, cubic symmetry, and a naked nucleocapsid, and Figure 2.4 describes DNA-containing viruses having envelopes or complex coats.

15

Ribonucleic acid (RNA)-containing viruses are presented in three figures: Figure 2.5, those with cubic capsid symmetry; Figure 2.6, those with helical symmetry; and Figure 2.7, those with capsid architecture that is either asymmetric or unknown. Commentaries follow on the viruses that have been definitely assigned to these groups. Also included are hepatitis viruses A and B and some other agents whose classification remains tentative.

THE NEW TAXONOMY

DNA Viruses

Parvoviridae (8). Originally named picodnaviruses to reflect their small size and DNA genome (9), the family of Parvoviridae now includes two named genera, *Parvovirus* and *Densovirus,* and a "probable genus," *Adenosatellovirus* (adeno-associated virus group). A typical member is adeno-associated satellite virus, several serotypes of which are indigenous to man. Reading from the left side of Figure 2.3, these are DNA-containing viruses that have cubic symmetry and a naked (unenveloped) nucleocapsid; during replication, capsid assembly takes place in the nucleus of the host cell.* Infectivity of these viruses is resistant to ether and other lipid solvents, the capsid has 32 capsomeres, the diameter of the virus particle is 18–26 nm, and the molecular weight of the nucleic acid is between 1.5 and 2.2×10^6. The capsomeres that form the outer layer of the nucleocapsid are each 2–4 nm in diameter.

The genus *Parvovirus* includes autonomously replicating members of the group that infect vertebrates: hamster osteolytic H viruses, latent rat viruses (Kilham rat virus, X14 virus), minute virus of mice (MVM), and parvoviruses of swine, cattle, cats, and other species. Members of the *Densovirus* genus are viruses of insects but are also capable of producing cytopathic effects in L cells (of vertebrates); they replicate autonomously. Adenosatelloviruses, however, are defective and cannot multiply in the absence of a replicating adenovirus that

---

* For the DNA viruses whose capsid assembly takes place in the nucleus (parvoviruses, papovaviruses, adenoviruses, and herpesviruses), a phase of replication (i.e., viral protein synthesis) occurs in the cytoplasm; thus viral messenger RNA is found associated with polyribosomes.

16

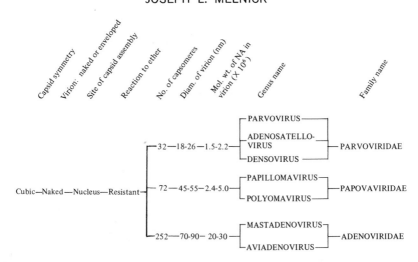

Figure 2.3. DNA-containing viruses with cubic symmetry and naked nucleocapsid. (NA = nucleic acid.)

serves as a "helper virus." Herpesvirus can act as a partial helper; in cells coinfected with herpesvirus, infectious satellite DNA and capsid proteins are produced, but they are not assembled into satellite virions.

Members of the family Parvoviridae are the only DNA-containing viruses of vertebrates whose DNA genome is single-stranded within the virion; all the others shown in Figures 2.3 and 2.4 have double-stranded DNA. In adenosatelloviruses and densoviruses, separate virions contain single strands of positive or negative DNA; these strands are complementary, and when isolated from the virion shells, they come together to form a double strand. However, in members of the genus *Parvovirus* (of which Kilham's rat virus is the type species), the DNA in the virion is a positive strand only. Members of this genus show preference for actively dividing cells, have been shown to be transmissible transplacentally, and are receiving attention for their special disease potential in fetuses and neonates (10,11).

Accumulating data clearly show that hepatitis virus type B has a number of important properties similar to those of representative members of the parvovirus family. The 42-nm "Dane" particle found in the serum of hepatitis B virus-infected persons is now recognized as the virus causing this disease. The morphology, nucleic acid type, and nucleic acid strandedness of the Dane particles place them in a class

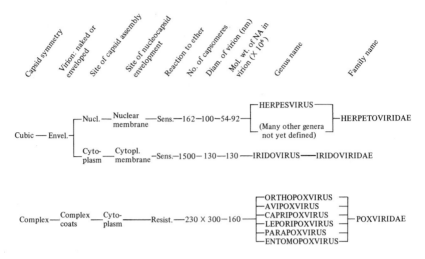

Figure 2.4.   DNA-containing viruses with envelopes or complex coats.

by themselves, unrelated to any other known viruses. However, the 26-nm central core (HBcAg, hepatitis B core antigen) that can be released from Dane particles shares many biochemical and biophysical properties with several members of the parvovirus family.

In addition to similarities in site of maturation, morphology, and size, core particles and members of the parvovirus family contain three polypeptides with similar molecular weights (12). Moreover, they share a similar sedimentation coefficient and DNA molecular weight; the buoyant density of core particles resembles that of defective adeno-satellite viruses. However, core particles and parvoviruses seem to differ in two major characteristics. First, the DNA of core particles has been reported to be double-stranded (13–15), whereas parvovirus virions contain single-stranded DNA (16). One should note, however, that the DNA of one group of parvoviruses, namely, the adenosatellite viruses, or adenosatelloviruses, had been variously reported as double-stranded or single-stranded, that this seeming contradiction was only resolved when positive and negative single strands of DNA were found within different adenosatellite virions, and that the strands on extraction from the virions united to form double-stranded DNA (17,18). As yet, no report has appeared on the isolation of the DNA of hepatitis core particles under conditions that would prevent reannealment of positive and negative single DNA strands, as has been done with the adenosatellite viruses.

18

Another problem to be reconciled is the fact that hepatitis core particles exist as units within the nuclei of infected hepatocytes in which they are manufactured, but when they circulate in the blood, they are present within a shell, the entire unit considered now to be a Dane particle or the virus. Hepatitis core particles (devoid of an outer shell) appear to be more like members of the parvovirus family than any other known virus group. However, other observations suggest that the two differ, and eventually hepatitis B virus may be placed in a virus group that has not previously been recognized.

Papovaviridae (19,20) (see Chapter 11). These relatively small, ether-resistant viruses contain double-stranded DNA in circular form. Many are unusually heat-stable, surviving temperatures that inactivate most viruses. Representative members that infect human beings are the papilloma, or wart, virus and SV40-like viruses, such as JC virus, which has been isolated from the brain tissue of patients having progressive multifocal leukoencephalopathy (PML), or BK virus, which has been isolated from the urine of immunosuppressed recipients of renal transplants (21). Other members include papillomaviruses of several vertebrate species, polyoma and K viruses of mice, and vacuolating viruses of monkeys (SV40) and of rabbits. These viruses have relatively slow growth cycles characterized by replication within the nucleus. Papovaviruses produce latent and chronic infections in their natural hosts; all are tumorigenic in at least some animal host species. The genome, a single cyclic molecule of double-stranded DNA, integrates into cellular chromosomes of transformed cells. The capsid antigens of JC and BK viruses are unique, but the tumor antigen induced by each cross-reacts antigenically with that induced by SV40.

When SV40 and adenoviruses replicate together within the same cell, they may interact to form various kinds of SV40-adenovirus "hybrid" virus particles, in which a portion of the SV40 genome is covalently linked to incomplete or complete adenovirus DNA and is carried within an adenovirus capsid (see p. 36).

Adenoviridae (22). Among these medium-sized viruses, at least 33 serotypes infect man, and distinct serotypes exist for a number of other species. The virion is a nonenveloped isometric particle with 252 capsomeres, each 7–9 nm in diameter. Vertex capsomeres are antigenically distinct from the others and carry one or two filamentous projections.

The adenovirus genome is a single linear molecule of double-stranded DNA.

Adenoviruses have a predilection for mucous membranes and may persist for years in lymphoid tissue. Some of the adenoviruses cause acute respiratory diseases, febrile catarrhs, pharyngitis, and conjunctivitis. Human adenoviruses rarely cause disease in laboratory animals, but certain serotypes produce tumors in newborn hamsters. Common antigens are shared by all mammalian adenoviruses, which now have been classified as members of the *Mastadenovirus* genus; these antigens are different from the corresponding antigens of the members of the genus *Aviadenovirus*.

Herpetoviridae (23) (see Chapters 9 and 10). Herpesviruses are a heterogeneous group of viruses identified by their structure. As indicated in Figure 2.5, the particle consists of a DNA-containing core enclosed by an icosahedral capsid with 162 hollow cylindrical capsomeres. A lipid membrane containing virus-specific proteins surrounds the inner structures. The double-stranded DNA of various herpesviruses differs considerably in size (50 to 135×10^6 molecular weight), cytosine + guanine content (44% to 74%), and structural complexity. Herpes simplex virus has a complex structural organization, including in the genome a terminally redundant section and internal inverted repetitions of sequences present at both ends of the DNA molecule with a long and short unique sequence region. Other members of the herpesvirus family may have simpler structural organization of the genome. The DNA of herpesviruses is sufficiently large to code for 80 to 100 proteins, of which about 50 have been observed. As many as 30 of these may be structural proteins of the virus particle, whereas others may be the virus-induced enzymes, including thymidine kinase, DNA polymerase, and DNase.

Herpesviruses are noteworthy for their ability to establish latent or persistent infections or both. Latent infections may last for the lifetime of the host, even in the presence of circulating antibodies. Special interest has been generated by the association of Epstein-Barr herpesvirus (EBV) with human Burkitt lymphoma and nasopharyngeal carcinoma (see Chapter 10) and by the possible role of the genital herpesvirus, HSV type 2, in cancer of the uterine cervix. Several simian herpesviruses have been shown to be oncogenic in experimentally infected animals. Infections of heterologous species are often very serious; examples are the fatal infection of man by one of the simian herpesviruses, the so-called B virus, and the infection of cattle by swine pseudorabies virus. Human diseases caused by herpesviruses include oral and genital lesions, chickenpox and shingles due to varicella-zoster virus, cytomegalic inclusion disease, and infectious mononucleosis.

Clearly, the members of the herpetovirus family are numerous; however, classification has not been agreed on and, to date, only one genus has been formally recognized, *Herpesvirus,* of which HSV type 1 is the type species. A classification of herpesviruses into subgroups A and B, based on whether the virus was cell-free or cell-associated, proved useful before the days of molecular virology (24).

Iridoviridae (25,26). The best known members of this family are the members of the insect iridescent virus group (e.g., *Tipula* iridescent virus), now placed in the genus *Iridovirus*. However, other important viruses that are considered probable members of this family include African swine fever virus and a large number of viruses of frogs and fish. No human iridovirus is known. Vertebrate iridoviruses are enveloped; iridoviruses that infect insects contain a lipid fraction in the virion but do not have an envelope as such. The genome is a single large molecule of double-stranded DNA.

Poxviridae (27). These large viruses are brick-shaped or ovoid, with a complex virion structure: an external coat contains lipid and tubular or globular protein structures; the coat encloses one or two lateral bodies and an internal body (core) that contains the genome. The virion contains more than 30 structural proteins and several viral enzymes, including a DNA-dependent RNA polymerase. The genome consists of a single molecule of double-stranded DNA. Genetic recombination occurs within genera; nongenetic reactivation occurs both within and between genera of the poxviruses that infect vertebrates. Most poxviruses of vertebrates share at least one antigen; members of each genus of vertebrate poxviruses have additional antigens in common. This is the major DNA-containing virus family whose members replicate entirely within the cytoplasm; a number of them produce intracytoplasmic inclusion bodies (type B—viral factory; and type A—cytoplasmic accumulation). The family has been divided into six

21

genera; the genus *Orthopoxvirus* includes the poxviruses of man. This genus produces a hemagglutinin, separate from the virion; the hemagglutinin is serologically specific and is a lipid-rich pleomorphic particle 50–65 nm in diameter.

RNA Viruses

Picornaviridae (28). Picornaviridae are shown at the top of Figure 2.5. Members of this family, which comprises the smallest of the viruses with RNA genomes, exist in at least two genera and in several hundred species. At least 70 members of the *Enterovirus* genus are known to infect man; these include polioviruses, coxsackieviruses, echoviruses, and, in recent years, new enterovirus serotypes that are assigned sequential numbers, enterovirus-68 . . . and the like, rather than being placed in ill-defined subgroups. Well over 100 viruses infecting human beings belong to the genus *Rhinovirus*. Large numbers of agents from both of these genera also exist for other species.

The picornavirus genome is one piece of linear, single-stranded RNA of low molecular weight (about 2.5×10^6). The RNA is infectious and serves as its own messenger for protein translation. The

Figure 2.5. RNA-containing viruses with cubic capsid symmetry.

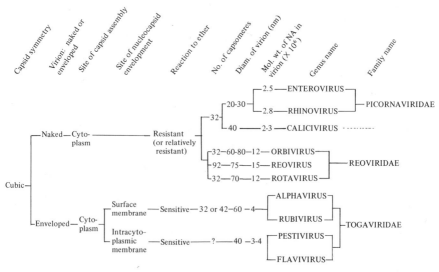

enteroviruses are acid-stable and have a buoyant density in cesium chloride of about 1.34 gm/cm³; the rhinoviruses, in contrast, are acid-labile and have a higher buoyant density, about 1.4 gm/cm³.

Several other groups of viruses have been considered for possible membership in the picornavirus family, including the *ribophages* (RNA-containing bacteriophages resembling picornaviruses in a number of properties) and the *caliciviruses* (the viruses causing swine vesicular exanthema and a respiratory disease of cats, as well as several viruses that infect sea lions). Neither ribophages nor caliciviruses have as yet been firmly placed as picornaviruses, for there are several important differences, as shown in Table 2.1 (28). In the current determinations of the ICTV (3), ribophages have not been included in the family, and caliciviruses are only tentatively included under the designation of a "possible genus." The caliciviruses have distinctive staining and have only one structural polypeptide rather than several. However, an important reason for their being considered as picornaviruses is the manner by which they replicate or their "genome strategy." To the extent that this strategy is known, for the caliciviruses, it closely resembles that of enteroviruses; in contrast, the genome strategy of the ribophages is dissimilar.

TABLE 2.1. *Properties Differentiating Caliciviruses and Ribophages from the Picornaviruses*

Property	Picornaviruses	Caliciviruses	Ribophages
Virus particles			
Diameter	24–30 nm	35–40 nm	25nm
Sedimentation value	140–160S	170–180S	80S
Molecular weight	8.5×10^6	$>10 \times 10^7$	3.6×10^6
RNA			
Sedimentation value	33–35S	37S	27S
Molecular weight	2.5×10^6	3.0×10^6	1.1×10^6
Polypeptides	4 major	1 major	1 major
	1 minor	1 minor	1 minor
Replication	Translation into large protein precursor that is subsequently cleaved (monocistronlike messenger)	Resemble picornaviruses	Translation into three major polypeptides (polycistronic messenger)

Cooper (29) has suggested that genome strategy may be a substantial factor in indicating relatedness and perhaps a more fundamental one than, for example, capsid structure, which might be more readily altered by mutational or environmental manipulation. Whereas a common strategy of replication naturally includes a common particle structure, the criterion of strategy may be useful to differentiate certain viruses that have attained some mutual similarity in structure via different evolutionary pathways. This cannot be the only criterion for degrees of relatedness, since viruses with similar genome strategies but no other properties in common could have arrived at this similarity by converging in the course of evolution. But viruses with distinct strategies would clearly be expected to have arisen through distinct phylogeny and thus would not be placed in the same family. As applied to the picornavirus "candidate genera," this distinction would mean that ribophages, with widely different genome strategies, would clearly be excluded from the Picornaviridae. In contrast, caliciviruses would be more likely to belong to the family by this criterion, despite the morphological and structural differences from viruses already placed as Picornaviridae.

Diseases caused by picornaviruses range from severe paralysis (paralytic poliomyelitis) to aseptic meningitis, pleurodynia, myocarditis, skin rashes, and common colds; inapparent infection is common. Different viruses may produce the same syndrome; on the other hand, the same picornavirus may cause more than a single syndrome.

After decades of investigation, it appears that hepatitis A virus is an enterovirus, although more data are needed to settle the issue (30). Although the nucleic acid type for hepatitis A virus has not been conclusively determined, the evidence strongly suggests that it is RNA. Hepatitis A virus also resembles enteroviruses in its 27-nm size and spherical shape, its density in cesium chloride (1.34 gm/cm^3), its location in the cell cytoplasm, and its stability to ether and to acid pH. Both hepatitis A virus and enteroviruses are labile at $100°C$ temperature, but hepatitis A virus is more consistently stable at $60°C$ heat, whereas enteroviruses vary in this regard. The agent is clearly distinguished from the rhinoviruses by its acid stability; from the hepatitis B virus (i.e., the Dane particle) by size, nucleic acid type, and location within the cell; and from HBsAg, which is a smaller particle without any demonstrable nucleic acid (see Chapter 4). Hepatitis A virus can be distinguished from the parvoviruses on the basis of size, nucleic acid type, and location of parvoviruses in the cell nucleus.

Reoviridae (31). Members of this virus family share a property unique among the RNA-containing viruses of vertebrates, in that the genome is composed of double-stranded rather than single-stranded RNA and consists of several segments. The capsid has a double shell; the structure of the outer capsid layer is indistinct, but icosahedral symmetry has been demonstrated in the inner capsid layers of all three recognized groups of reoviruses that infect vertebrates, that is, the genera *Reovirus* and *Orbivirus* and the probable genus *Rotavirus*. Members of the genus *Reovirus* have been thought to have 92 capsomeres (but the number is being restudied), whereas each of the other two groups has 32 capsomeres. The capsomeres of the orbiviruses are unusually large (10–15 nm wide) and appear ring-shaped. The human reoviruses are found in the enteric tract, but their association with disease is not clear; members of this genus recovered from lower animals are similar to those of man. The *Orbivirus* genus includes viruses that infect both vertebrates and invertebrates; some have been considered to be arboviruses (see below, p. 26). Several have been recovered only from insects. The diseases caused by orbiviruses include blue-tongue, African horse sickness, Colorado tick fever of man, and epizootic hemorrhagic disease of deer. The members of the probable genus, *Rotavirus,* that infect human beings are increasingly recognized as major pathogens of nonbacterial infantile diarrhea; the gastroenteritis syndrome is clinically more severe and of longer duration than the illness caused by the 27-nm "Norwalk agent" and occurs in sporadic rather than epidemic form. It is one of the most common childhood illnesses throughout the world and is a leading cause of death in underdeveloped countries. Other members of this antigenically interrelated rotavirus group include calf diarrhea virus, the virus of epizootic diarrhea of infant mice, SA11 rotavirus of monkeys, and similar viruses from swine and other species. Much of the initial study of the rotaviruses was accomplished by use of electron microscopy, and thus far, isolation of the human members of the group has not been achieved in cell culture. The virus does replicate in fetal intestinal organ cultures, and antibody to the virus can be assayed by indirect immunofluorescence (32).

In addition to the members of Reoviridae that infect vertebrates, at least two other groups within the family are considered genera, although they do not yet have official generic names: the *cytoplasmic polyhedrosis group of insect viruses* and the *plant reovirus group.*

Togaviridae (7) (see Chapters 6 and 7). Members of this family include most arboviruses* of antigenic groups A and B, now classed in the genus *Alphavirus* (group A), the genus *Flavivirus* (group B), and in newly designated genera that include nonarbo togaviruses, rubella (*Rubivirus*), and the mucosal disease virus group (*Pestivirus*). The virions are spherical, 40–70 nm in diameter, and have a lipoprotein envelope with lipid and virus-specified glycopeptide tightly applied to an icosahedral nucleocapsid. The genome is a single molecule of single-stranded RNA. The alphaviruses and flaviviruses include many of the major human arboviral pathogens: the viruses of Venezuelan, eastern, and western equine encephalitis are alphaviruses, and the viruses of yellow fever, dengue, Japanese encephalitis, St. Louis encephalitis, Omsk hemorrhagic fever, and Russian spring-summer encephalitis are flaviviruses. Rubella virus thus far is the only member of the genus *Rubivirus:* equine arteritis virus may also be a member of this genus. Members of the *Pestivirus* genus include bovine virus diarrhea (mucosal disease complex) virus, hog cholera (European swine fever) virus, border disease virus, and probably also lactic dehydrogenase virus of mice and simian hemorrhagic fever virus.

Orthomyxoviridae (34) (see Fig. 2.6). The orthomyxoviruses recognized to date are the influenza viruses, which may be spherical, elongated, or filamentous. For most members of the family, "spikes" pro-

---

* One important and well-known virus group name that does not appear in the diagrams is a category based on ecologic properties, the arbovirus group (33). The more than 350 arthropod-borne viruses survive through a complex cycle involving vertebrate hosts and arthropods that serve as vectors, transmitting the viruses by their bites. This grouping, based on transmission, remains a useful one despite the wide diversity of its members with regard to properties of the virion. Most arboviruses now have been sufficiently well characterized to permit their taxonomic placement. Their classic serologic interrelationships previously delineated by arbovirologists have been found to be paralleled by morphological similarities, and these serologic relationships have tremendously speeded the taxonomic placement process by providing clues on where to look. Once some of the members of a serologic group have been characterized in terms of biophysical and biochemical properties, attention of taxonomists can be focused on the antigenic relatives of those members. Arboviruses now are included in a number of families, chiefly Togaviridae, Bunyaviridae, Rhabdoviridae, Arenaviridae, and Reoviridae.

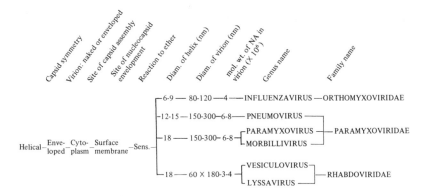

Figure 2.6.   RNA-containing viruses with helical symmetry.

ject from the surface of the envelope; these are glycosylated protein peplomers 10–14 nm long and 4 nm in diameter, consisting of two types: the hemagglutinin and the neuraminidase. The helically symmetric ribonucleoprotein capsid has RNA strands 6–9 nm in diameter (for influenza types A and B, 6 nm; for type C, 9 nm). The RNA is single-stranded, in six or seven segments. An RNA-dependent RNA polymerase is associated with purified virions. During replication, the helical nucleocapsid is first detected in the nucleus, whereas the hemagglutinin and neuraminidase are formed in the cytoplasm. The virus matures by budding at the cell surface membrane.

The *Influenzavirus* genus includes viruses of type A and type B; type C is considered only a "probable genus" (see Chapter 5). Antigenic variation is common, particularly among members of type A. Recombination occurs with high frequency within species but not between types or genera. Type A influenza viruses include agents of human, equine, and swine influenza and of fowl plague. For types B and C, only human strains are known.

Paramyxoviridae (3). Usually spherical, these virions may also be pleomorphic. They are 150 nm or more in diameter, and filamentous forms may be several micrometers long. On the lipid bilayer envelope are surface projections. The RNA is single-stranded and is in unsegmented, linear form; the helical nucleocapsid is 12–18 nm in diameter. Virions are formed in the cytoplasm by budding from the plasma membrane. Infectivity is sensitive to ether, acid, and heat, but paramyxo-

27

viruses are resistant to dactinomycin. The genera include *Paramyxovirus* (parainfluenza viruses, mumps virus, Newcastle disease virus, and Yucaipa and other avian paramyxoviruses), *Morbillivirus* (the virus of measles, canine distemper, rinderpest and pest de petits ruminant), and *Pneumovirus* (respiratory syncytial viruses of man and of cattle, and pneumonia virus of mice). Members of the genus *Paramyxovirus* have both hemagglutinin and neuraminidase in the virion. *Morbillivirus* members have hemagglutinin in the viral envelope but not neuraminidase, whereas for members of the genus *Pneumovirus* the virions contain neither hemagglutinin nor neuraminidase. Members of Paramyxoviridae are genetically stable, and genetic recombination does not occur.

Rhabdoviridae (3). Members of this family have enveloped virions that are rod-shaped, resembling a bullet (with one end rounded and the other flattened) or bacilliform. Enclosed within the lipoprotein envelope and membrane protein is the long tubular nucleocapsid with helical symmetry. The genome of single-stranded RNA is in unsegmented, linear form. Members of some genera multiply in arthropods as well as in vertebrates or higher plants; others multiply only in insects. Infectivity is sensitive to ether, acid, and heat. The genera that infect vertebrates are *Lyssavirus* (including rabies virus, Duvenhage virus, and Mokola virus that infect man; Lagos bat virus; and several agents thus far isolated only from insects) and *Vesiculovirus,* including vesicular stomatitis virus and a number of antigenically interrelated viruses. Among the vesiculoviruses are Chandipura virus (from man), Flanders-Hart Park virus (of mosquitoes and birds), Kern Canyon bat virus, Piry opossum virus, and probably also bovine ephemeral fever virus, Mt. Elgon bat virus, Egtved virus (viral hemorrhagic septicemia virus) of trout, and several other fish viruses. In addition, a genus, *Sigmavirus,* has been established for the rhabdoviruses of *Drosophila,* and there also are numerous rhabdoviruses of plants. Marburg virus, a simian virus highly pathogenic for man, is rhabdoviruslike in most characteristics but has extremely elongated forms.

Retroviridae (35,36) (see Fig. 2.7). Much remains to be settled concerning this family, although much progress has been made. The members include not only the RNA tumor viruses ("oncornaviruses," "leukoviruses") (see Chapter 12), that are now assigned to a sub-

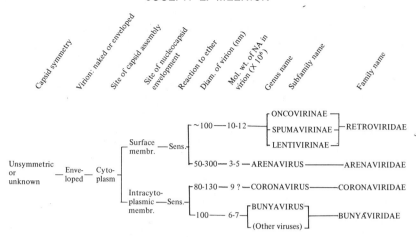

Figure 2.7. RNA-containing viruses with architecture unsymmetric or unknown.

family, Oncovirinae, but also the slow viruses of the maedi/visna group (now subfamily Lentivirinae) and the foamy virus group of agents that form syncytia in cell cultures (now assigned to the subfamily Spumavirinae).

Members of the family characteristically have a reverse transcriptase (RNA-dependent DNA polymerase) within the virion. For the most thoroughly studied members, the lipoprotein envelope encloses an inner shell with icosahedral symmetry and a central core or nucleocapsid with helical symmetry. The genome is a single molecule of single-stranded RNA that dissociates readily into two or three pieces. Infectivity is ether-, acid-, and heat-sensitive. Replication of the viral RNA involves a DNA provirus that is integrated into host cellular DNA.

With Oncovirinae, all normal cells of several animal species contain integrated copies of genes of the endogenous species of oncovirus. The oncovirus genes may not be expressed but can be activated by physical and chemical agents, by superinfection with other oncoviruses, and even by herpesviruses. According to their host range in tissue culture, oncoviruses fall into three categories: (a) ecotropic—capable of growth in cells of the natural host; (b) xenotropic—capable of growth only in cells of a different species; and (c) amphotropic—capable of growth in both autologous and heterologous cells. According to certain morphological, antigenic, and enzymatic differences, oncoviruses also have been

29

divided into A, B, and C (and possibly D) types of viruses. Most current schemes for classifying oncoviruses utilize Bernhard's classification of A, B, and C particles. A-type particles are double-shelled and have an electron-lucent center. B-type particles have eccentric cores, and C-type particles have central cores.

With some exceptions, oncoviruses fall into host-species-specific groups of agents, inducing either leukemias or sarcomas, that is, leukemia-sarcoma complexes of avian, murine, feline, and hamster oncoviruses; other groups are murine mammary tumor virus and primate oncoviruses. One primate oncovirus is the monkey mammary tumor virus (MoMTV, previously termed Mason-Pfizer virus) from a rhesus monkey tumor; it may be a D-type agent, the particles generally resembling B or C particles but having a distinct configuration with an outer unit membrane and an inner tubular or cylindrical nucleoid.

Recent evidence indicates that all strains of mice studied to date contain C-type viruses that are xenotropic (i.e., replicate in cells of other species but not in the homologous murine cells).

The ICTV (3) has designated two genus-level groups within this subfamily, although generic names have not been assigned. These are (a) the type C oncovirus group (including mammalian, avian, and reptilian subgenus groupings of type C viruses, as well as a number of ungrouped species); and (b) the type B oncovirus group in which the virus core is located eccentrically within the extracellular virion. These include the mouse mammary tumor virus and similar viruses from guinea pig and perhaps other species.

The members of the subfamily Spumavirinae, "foamy viruses," do not induce tumors or cellular transformation but cause persistent asymptomatic infections in natural and experimental host animals. They have perhaps been best known because they induce syncytia in cell cultures being prepared for cultivation of other viruses. Foamy or syncytial viruses are known for a number of mammalian species, perhaps including man.

The slow viruses of the maedi/visna group that have been placed in the subfamily Lentivirinae of the Retroviridae family are morphologically and chemically like other members of the family but do not induce tumors. A nucleoid or core forms immediately beneath the viral envelope during budding from the cellular plasma membrane, and particles lack a second internal viral membrane. Natural infections are known only in sheep. Visna virus causes panleukoencephalitis and infects all the organs of the infected sheep. Pathological changes, however, are confined chiefly to the brain, lungs, and reticuloendothelial system. The incubation period is long, and virus can be recovered from the animal as long as four years after inoculation. Serologically related viruses (variously designated in different countries as maedi, progressive pneumonia, or Zwoegerziekte viruses) cause interstitial pneumonitis.

Arenaviridae. Members of Arenaviridae (37) have spherical or pleomorphic virions with a dense lipid bilayer membrane bearing surface projections. Within the virion core are electron-dense RNA-containing granules about 20–30 nm in diameter that resemble ribosomes in size, shape, and density. The RNA is single-stranded, and consists of four large and one to three small segments. Virions are formed by budding from the surface membrane. Viral RNA is probably transcribed by virion polymerase into complementary RNA, which probably acts as messenger RNA (mRNA). Infectivity is sensitive to ether, acid, and heat. Most member viruses have a single restricted rodent host in which persistent infection occurs accompanied by viremia, viruria, or both. Spread to other mammals and to man occurs but is unusual. The Arenaviridae include lymphocytic choriomeningitis virus, which infects mice but may spread to man; Lassa virus; and members of the Tacaribe complex, which comprises Junin and Machupo viruses of South American hemorrhagic fevers, Pichinde virus, and several other viral agents that have been isolated from arthropods but for which natural transmission cycles involving arthropod vectors have not been demonstrated.

Coronaviridae (38). The family is named for unique petal-shaped or club-shaped peplomers that project from the envelope. In negatively stained electron micrographs, these projections form a fringe resembling the solar corona. The interior structure of the virion is not fully understood but probably is a loosely wound, helically symmetric nucleocapsid. The genome consists of one large molecule of single-stranded RNA. Infectivity is sensitive to ether, acid, and heat. Nucleocapsids develop in the cytoplasm and mature by budding through intracytoplasmic membranes. Several serotypes of human coronaviruses have been isolated through the use of human embryonic tracheal and nasal organ cultures. There are distinct coronaviruses that infect a number of animal species: avian infectious bronchitis virus, turkey

31

blue-comb disease virus, mouse hepatitis virus, porcine transmissible gastroenteritis virus and hemagglutinating encephalitis virus, calf neonatal diarrhea coronavirus, and at least two rat coronaviruses.

Bunyaviridae (39). This family is the largest and most recently recognized taxonomic grouping assigned to an antigenically interrelated set of arboviruses. The family has at least 150 members, more than 85 belonging to the Bunyamwera supergroup of arboviruses, which consists of 10 serologically cross-related groups and several ungrouped arboviruses. The virions are spherical and have surface projections on their unit-membrane envelope that may be randomly placed or clustered in arrays with icosahedral symmetry; the envelope contains at least one virus-specified glycopeptide. The internal ribonucleoprotein is helically wound and symmetric, with long strands 2–2.5 nm broad. The single-stranded RNA is probably in three circular segments. The virions develop in the cytoplasm and mature by budding through intracytoplasmic membranes into smooth-surfaced vesicles in or near the Golgi region. Infectivity is sensitive to ether, acid, and heat. Virus particles hemagglutinate. Members of the family produce a number of important diseases of man and of domestic animals, for example, California encephalitis, Crimean hemorrhagic fever, sandfly fever, Rift Valley fever, and Nairobi sheep disease. Most Bunyaviridae members are mosquito-transmitted, but some are tick-borne. In addition to the viruses already assigned to the genus *Bunyavirus* (confined at present to members of the serologic Bunyamwera supergroup), a number of other arboviruses are generally similar to bunyaviruses in most properties but are not serologically related. These agents are considered possible members of the family, perhaps in different genera yet to be designated. The largest antigenic grouping among these is the Uukuniemi group, antigenically unrelated to *Bunyavirus* and transmitted by ticks.

OTHER CLASSIFICATION SCHEMES

Although vertebrate virus families can be diagrammed and discussed conveniently in an organizational arrangement wherein the primary separation is based on the type of nucleic acid genome, the known data about groups of viruses can be organized in other ways. One interesting approach to virus classification relates to the induction or carriage of polymerases, the enzymes essential for replication of the viral genes. As shown in Table 2.2, among viruses with an RNA genome, the viruses with fewer genes (picornaviruses, togaviruses) induce, but do not carry, RNA-dependent RNA polymerase, whereas the medium-sized RNA viruses carry that enzyme. The retroviruses are distinctive in that the enzyme they carry is an RNA-dependent DNA polymerase (reverse transcriptase). Among the DNA-containing viruses, the small- and medium-sized viruses induce DNA-dependent DNA polymerase; only the large poxviruses both induce this enzyme and carry DNA-dependent RNA polymerase.

Another major difference among viruses is the presence or absence of a lipid-containing envelope. Matthews (40) developed an interest-

TABLE 2.2. *Use of Polymerases for Classification of Animal Viruses*

Virus Family	Example	Approximate Number of Genes	Polymerase
RNA Viruses			
Picornaviridae	Poliovirus	12 ⎫	Induce RNA-dependent
Togaviridae	Japanese B encephalitis	15 ⎭	RNA polymerase
Reoviridae	Reovirus	40 ⎫	
Orthomyxoviridae	Influenza virus	15 ⎪	
Paramyxoviridae	Measles virus	30 ⎪	
Rhabdoviridae	Vesicular stomatitis virus	20 ⎪	Carry RNA-dependent
Arenaviridae	Lymphocytic choriomeningitis virus	15 ⎬	RNA polymerase
Bunyaviridae	Bunyamwera virus	15 ⎪	
Coronaviridae	Human upper respiratory illness virus	30 ⎭	
Retroviridae	C-type oncovirus	50	Carry RNA-dependent DNA polymerase
DNA Viruses			
Parvoviridae	Adenosatellite virus	7 ⎫	
Papovaviridae			
Polyomavirus	Polyoma virus	7 ⎪	Induce DNA-dependent
'Papillomavirus	Human wart virus	13 ⎬	DNA polymerase
Adenoviridae	Human adenovirus	50 ⎪	
Herpetoviridae	Herpes simplex virus	180 ⎭	
Poxviridae	Vaccinia virus	400	Induce DNA-dependent DNA polymerase and carry DNA-dependent RNA polymerase

ing and potentially useful classification based in part on the viral membrane. His system is based on the relationship between the size of the viral genome and the size of the entire virion (dry mass or particle volume). When that relationship is determined, viruses fall into two classes: those with and those without envelopes. The enveloped viruses have a ratio of nucleic acid molecular weight to whole virus anhydrous weight of 1 : 40 and to whole virus volume of 1 : 0.2 (Table 2.3). These ratios of genome to weight and volume are similar to those of prokaryotic cells.

In contrast, nonenveloped viruses have a ratio of nucleic acid to whole virus dry weight of 1 : 4 and to whole virus volume of 1 : 0.01, decidedly different from both the enveloped viruses and the prokaryotic cells. Another manner in which these two classes of viruses differ is in the reaction of enveloped viruses to freezing. Wallis and I (41) found that enveloped viruses required the presence of the same additives to preserve their infectivity that are required to preserve the membranes and viability of animal cells, whereas such additives were not required to stabilize nonenveloped viruses.

Further subdivisions could be made based on the size of the nucleic acid and its mode of replication. For the enveloped viruses (Fig. 2.8), those with large genomes have double-stranded DNA, and those with smaller genomes have single-stranded RNA. For the nonenveloped viruses (Fig. 2.9), those with genomes above a certain size have double-stranded nucleic acid (DNA or RNA), whereas those below that size have single-stranded nucleic acid (DNA or RNA).

The striking difference between the two virus classes can be brought into focus by comparing influenza virus and poliovirus. Influenza virus is only 1/1000th the size of *E. coli,* but for both virus and bacterium, the values determined for dry mass and particle volume per unit of nucleic acid are approximately the same. When influenza virus is com-

TABLE 2.3. *Comparison of Enveloped and Nonenveloped Viruses: Virus and Genome Size*

	Ratio of Molecular Weight of Nucleic Acid to:	
Class of Virus	Whole Virion Dry Weight	Whole Virion Volume
Enveloped	1 : 40	1 : 0.2
Nonenveloped	1 : 4	1 : 0.01

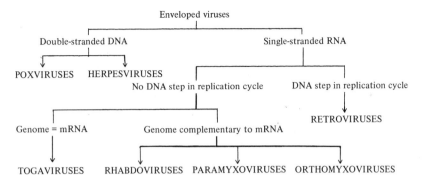

Figure 2.8.   Matthews' classification system: enveloped viruses.

pared with poliovirus, one finds that the former has about the same amount of RNA as the latter, but the anhydrous mass of influenza virus is 10 times that of poliovirus, and the volume of whole influenza virus is 20 to 30 times greater (Table 2.4). Matthews has suggested that a primary division of viruses in these two classes may have more predictive value than the current schemes considered earlier in this chapter and might correspond more nearly to viral evolution.

EMERGING PROBLEMS IN VIRUS CLASSIFICATION

Some of the present and developing problems that viral taxonomists will have to meet are those posed by the recently discovered forms of life called viroids, viral hybrids (between unrelated viruses), pseudovirions, and recombinant DNA.

Figure 2.9.   Matthews' classification system: nonenveloped, geometric viruses.

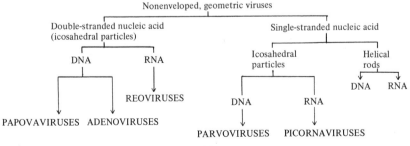

TABLE 2.4. *Comparison of Influenza Virus and Poliovirus*

Virus	Molecular Wt. Nucleic Acid (A)	Whole Virion Dry Mass (B)	Whole Virion Volume (C)	Ratios A/B	Ratios A/C
Influenza virus	$\sim 4 \times 10^6$	$\sim 2 \times 10^8$	$\sim 8 \times 10^5$	1 : 50	1 : 0.2
Poliovirus	2.6×10^6	8.5×10^6	1.2×10^4	1 : 3	1 : 0.005

Viroids. These organisms constitute a recently discovered class of infectious agents smaller than viruses. They are known to cause several diseases of plants (e.g., potato spindle tuber disease) and may ultimately be found to cause disease in man and higher animals. For example, the agent of scrapie disease of sheep, one of the puzzling "slow viruses" not yet placed taxonomically, may, according to recent findings, prove to be a viroid (see Chapter 13). Viroids exhibit the characteristics of nucleic acids in crude extracts, that is, they are insensitive to heat and to organic solvents but are sensitive to nucleases. They do not appear to have a protein coat. Viroids known at present consist solely of a short strand of RNA with a molecular weight of 75,000 to 100,000. ,

Virus Hybrids. The fact that virus hybrids can exist in nature should be more widely recognized. If SV40 had not already been known as a virus before the discovery of SV40-adenovirus "hybrid" particles, these particles would have presented viral taxonomists with a confusing puzzle. In these hybrid particles, portions of SV40 genome material are covalently linked to adenovirus genetic material encased within an adenovirus coat. They would, therefore, have seemed to be a new and strange virus that reacted antigenically as an adenovirus (of the serotype from which the coat had been derived) but that had many properties altogether unlike an adenovirus when grown in cultures.

Two types of adenovirus-SV40 hybrids have been detected. PARA-adenovirus populations consist of two types of particles: (a) a non-hybrid typical adenovirion and (b) a defective adenovirus-SV40 genome encased in an adenovirus capsid (PARA).

PARA can be transcapsidated from one adenovirus serotype to another. The second type of hybrid, the Ad2+ND viruses, consists of a series of nondefective adenovirus type 2 isolates carrying different amounts (5% to 44%) of the SV40 genome (Fig. 2.10).

A similar problem of identification and classification does, in fact, exist with respect to another type of particle found in some human adenovirus populations. That particle, termed MAC (monkey-adapting component), behaves somewhat like the PARA particle, permitting the true human adenovirus to replicate in monkey cell cultures. The particle, with a MAC genome and an adenovirus coat, does not contain any SV40 nucleic acid fragments, and its origin remains unknown.

Pseudovirion. Another viral form that is difficult to classify is the pseudovirion. During viral replication the capsid sometimes encloses host nucleic acid rather than viral nucleic acid. Such particles look like ordinary virus particles when observed by electron microscopy, but they do not replicate. Pseudovirions contain the "wrong" nucleic acid. For example, fragments of host-cell DNA (instead of viral DNA) may be incorporated into papovavirus capsids forming pseudovirion particles. This situation resembles the phenomenon of generalized transduction by bacteriophages (i.e., transfer of random portions of nucleic acid from donor to recipient bacterial cells). Hybridization studies also indicate the occurrence of covalent linkage of cell DNA segments into the circular DNA of papovaviruses during replication in cells infected at high multiplicity. This phenomenon is analogous to specialized transduction by bacteriophages (i.e., transfer by virus of a specific segment of donor bacterial cell DNA). Furthermore, under specialized experimental conditions, a DNA segment containing functional genes of λ bacteriophage has been incorporated into the circular DNA of papovavirus SV40. These findings open avenues for the study of possible transducing events in eukaryotic cells whereby functionally defined segments of genetic information might be transmitted from cell to cell. Pseudovirions present the taxonomist with problems based on natural events, but future laboratory manipulations will probably add to these problems of classification.

Recombinant DNA. Recently developed techniques allow DNA to be cleaved into specific pieces by use of enzymes from bacteria called restriction endonucleases. The distinct fragments have importance in two areas: (a) the physical mapping of genes in large, complicated DNA genomes and (b) genetic engineering. In addition to the over-

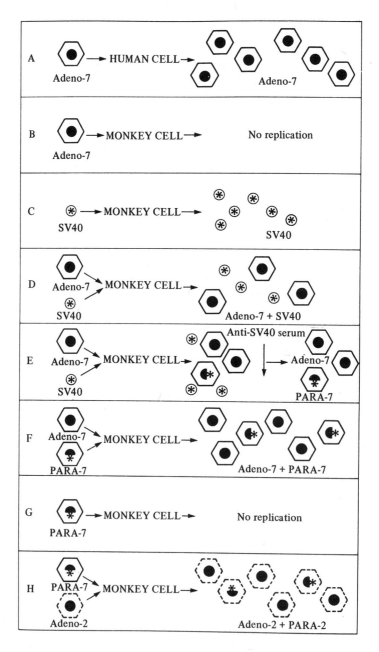

Figure 2.10. Mutual dependence between adeno particles and PARA in "hybrid" populations. The synthesis of SV40 tumor antigen by a "hybrid" requires the multiplication of both adeno particles and PARA. Adenovirus multiplies in human (A) but not in monkey kidney cells (B). SV40 repli-

riding concern for safety precautions to ensure that new genetic combinations thus produced do not result in new organisms with dangerous properties, virologists must also give attention to how the new recombinant organisms should be classified. Classification of these new forms of life needs to be developed in ways that will reflect their origin and relatedness to each other and to other living things.

As has been amply illustrated by this chapter, classification of viruses has progressed far since the early days, when most workers referred to viral taxonomy as "the science of calling names."

DISCUSSION

Question: Has the natural host of Marburg virus been established?

Dr. Melnick: Marburg virus disease is an acute febrile infection first recognized in 1967 during an epidemic among laboratory workers exposed to the infected tissues of imported African green monkeys (*Cercopithecus aethiops*) in Germany and Yugoslavia. Person-to-person nosocomial transmission occurred from the hospitalized primary cases to medical attendants. No further epidemics or sporadic cases have appeared among persons exposed to nonhuman primates in America or Europe. Nevertheless, serologic surveys have indicated that the virus is present in East Africa (Uganda and Kenya) and causes infections in monkeys and man. Two cases occurred in 1975 among tourists traveling through Africa. However, the natural host of the virus is unknown.

---

cates in monkey cells (C), where it also acts as a helper for adenovirus (D).

In the product of D, some adenovirus particles exist that contain defective SV40 DNA in their genome. As shown in E, such particles with the hybrid genome induce the formation of SV40 T antigen in the course of infection. Treating the product of the infected cells with anti-SV40 serum binds the SV40 particles, leaving a mixture of pure adenovirus-7 particles containing the combined adeno-SV40 DNA (E). The latter are called PARA because the particles aid in the replication of adenovirus in monkey cells (F). PARA particles are unable to replicate alone in monkey cells (G), but when a cell is infected with both pure adenovirus (type 2) and PARA of another antigenic adenovirus (type 7), transcapsidation occurs, and the PARA genome now acquires the coat of the helper adenovirus. This results in an SV40-adenovirus type 2 "hybrid" population containing both pure adeno-2 and PARA-2 particles (H). (From ref. 42.)

In 1976 an epidemic of hemorrhagic fever involving hundreds of cases, many of them fatal, occurred in southern Sudan and northern Zaire. Again, there was transmission from hospitalized primary cases to medical staff. The 1976 virus seems to be a member of the Marburg group since it is morphologically similar to the Marburg virus recognized previously, but it is antigenically distinct.

Question: Have any DNA-DNA hybridization studies been done to demonstrate a genetic relatedness between parvovirus DNA and DNA from the core of Dane particles?

Dr. Melnick: Not to my knowledge.

REFERENCES

1. Wildy P: Classification and Nomenclature of Viruses: First Report of the International Committee on Nomenclature of Viruses. *Monographs in Virology*. Basel, S Karger, vol 5, 1971.
2. Fenner F: The classification and nomenclature of viruses. Summary of results of meetings of the International Committee on Taxonomy of Viruses in Madrid, September 1975. *Intervirology* 6:1–12, 1975/76.
3. Fenner F: Classification and nomenclature of viruses: Second report of the International Committee on Taxonomy of Viruses. *Intervirology* 7:1–115, 1976.
4. Davis BD, Dulbecco R, Eisen HN, et al: *Microbiology*. New York, Hoeber, 1967.
5. Fenner F, White DO: *Medical Virology*, ed 2. New York, Academic Press, 1976.
6. Melnick JL: Taxonomy of viruses, 1976. *Prog Med Virol* 22:211–221, 1976.
7. Murphy FA: Taxonomy of vertebrate viruses, in *CRC Handbook of Microbiology*, 1977, in press.
8. Bachmann PA, Hoggan MD, Melnick JL, et al: Parvoviridae. *Intervirology* 5:83–92, 1975.
9. Mayor HD, Melnick JL: Small deoxyribonucleic acid-containing viruses (picodnavirus group). *Nature* 210:331–332, 1966.
10. Kilham L, Margolis G: Problems of human concern arising from animal models of intrauterine and neonatal infections due to viruses: A review. I. Introduction and virologic studies. *Prog Med Virol* 20:113–143, 1975.
11. Margolis G, Kilham L: Problems of human concern arising from animal models of intrauterine and neonatal infections due to viruses: A review. II. Pathological studies. *Prog Med Virol* 20:144–179, 1975.
12. Fields HA, Hollinger FB, Desmyter J, et al: Biochemical and biophysical properties of hepatitis B core particles derived from Dane particles and infected hepatocytes. *Intervirology* 8:336–350, 1977.
13. Hirschman SZ, Gerber M, Garfinkel E: DNA purified from naked

intranuclear particles of human liver infected with hepatitis B virus. *Nature* 251:540–542, 1974.

14. Overby LR, Hung PP, Mao JCH, et al: Rolling circular DNA associated with Dane particles in hepatitis B virus. *Nature* 225:84–85, 1975.

15. Robinson WS, Clayton DA, Greenman RL: DNA of a human hepatitis B virus candidate. *J Virol* 14:384–391, 1974.

16. Rose JA: Parvovirus reproduction, in *Comprehensive Virology*. New York, Plenum, 1974, vol 3, pp 1–61.

17. Mayor HD, Torikai K, Melnick JL, et al: Plus and minus single-stranded DNA separately encapsidated in adeno-associated satellite virions. *Science* 166:1280–1282, 1969.

18. Hoggan MD: Adenovirus-associated viruses. *Prog Med Virol* 12:211–239, 1970.

19. Melnick JL: Papova virus group. *Science* 135:1128–1130, 1962.

20. Melnick JL, Allison AC, Butel JS, et al: Papovaviridae. *Intervirology* 3:106–120, 1974.

21. Padgett BL, Walker DL: New human papovaviruses. *Prog Med Virol* 22:1–35, 1976.

22. Norrby E, Bartha A, Boulanger P, et al: Adenoviridae. *Intervirology* 7:117–125, 1976.

23. O'Callaghan DH, Randall CC: Molecular anatomy of herpesviruses: Recent studies. *Prog Med Virol* 22:152–210, 1976.

24. Melnick JL, Midulla M, Wimberly I, et al: A new member of the herpesvirus group isolated from South American marmosets. *J Immunol* 92:596–601, 1964.

25. Bellett AJD: The iridescent virus group. *Adv Virus Res* 13:225–246, 1968.

26. Kelly DC, Robertson JS: Icosahedral cytoplasmic deoxyviruses. *J Gen Virol* 20 (suppl):17–41, 1973.

27. Fenner F, Pereira HG, Porterfield JS, et al: Family and generic names for viruses approved by the International Committee on Taxonomy of Viruses, June 1974. *Intervirology* 3:193–198, 1974.

28. Melnick JL, Agol VI, Bachrach HL, et al: Picornaviridae. *Intervirology* 4:303–316, 1974.

29. Cooper PD: Towards a more profound basis for the classification of viruses. *Intervirology* 4:317–319, 1974.

30. Provost PF, Wolanski BS, Miller WJ, et al: Physical, chemical and morphologic dimensions of human hepatitis A virus strain CR 326. *Proc Soc Exp Biol Med* 148:532–539, 1975.

31. Joklik WK, et al: Reoviridae. *Intervirology*, to be published.

32. Holmes IH: Viral gastroenteritis. *Prog Med Virol* 24:1978, to be published.

33. Berge TO: *International Catalogue of Arboviruses Including Certain Other Viruses of Vertebrates*, DHEW publication no. (CDC) 75–8301. US Dept of Health, Education and Welfare, 1975.

34. Dowdle WR, Davenport FM, Fukumi H, et al: Orthomyxoviridae. *Intervirology* 5:245–251, 1975.

35. Dalton AJ, Melnick JL, Bauer H, et al: The case for a family of reverse transcriptase viruses: Retraviridae. *Intervirology* 4:201–206, 1974.

36. Vogt PK: The oncovirinae—a definition of the group, in *Report No. 1*

of the WHO Collaborating Centre for Collection and Evaluation of Data on Comparative Virology, Munich, 1976.

37.   Pfau CJ, Bergold GH, Casals J, et al: Arenaviruses. *Intervirology* 4: 207–213, 1974.

38.   Tyrrell DAJ, Almeida JD, Cunningham CH, et al: Coronaviridae. *Intervirology* 5:76–82, 1975.

39.   Porterfield JS, Casals J, Chumakov MP, et al: Bunyaviruses and Bunyaviridae. *Intervirology* 6:13–24, 1975/76.

40.   Matthews REF: A classification of virus groups based on the size of the particle in relation to genome size. *J Gen Virol* 27:135–149, 1975.

41.   Wallis C, Melnick JL: Stabilization of enveloped viruses by dimethyl sulfoxide. *J Virol* 2:953–954, 1968.

42.   Jawetz E, Melnick JL, Adelberg EA: *Review of Medical Microbiology,* ed 12. Los Altos, Calif, Lange Med Pub, 1976.

3

Virus-Cell Interactions

JAMES A. ROBB

Most encounters between humans and viruses do not produce clinical disease, although cellular disease can occur. Cellular disease is produced when suitable interactions occur between the virus and the host cell. Whether the virus-cell interaction produces clinical disease depends not only on the type of interaction occurring within individual cells but also on the reactions of each person to the virus-induced cellular disease. Here we describe various types of virus-cell interaction and give examples of human diseases produced by these interactions.

PHASES OF INFECTION

The phases of viral infection are shown schematically and in general terms in Figure 3.1. This schema is basically true for viruses containing either DNA or RNA genomes, regardless of whether they reproduce in the cytoplasm or nucleus.

1. *Attachment* requires, in most cases, virus-specific attachment sites on the outer surface of the cell, the plasma membrane. In general, this interaction involves at least one cellular and one viral glycoprotein. Lonberg-Holm and Philipson (1) have reviewed the attachment process in detail.

2. *Penetration* of the cell by the virus must then occur. This phase, accomplished by pinocytosis ("viropexis") and/or fusion of the membrane of enveloped viruses with the plasma membrane, remains controversial for many viruses (1). Pinocytosis involves the engulfment of the virus particle by the plasma membrane and the subsequent production of an intracellular membrane-bound vesicle containing the virus particle, a "sphere of destiny." Pinocytosis is the predominant mechanism of penetration for viruses that are not enveloped by a modified cellular membrane. Scheid and Choppin (2) have suggested that the predominant mechanism used by enveloped viruses for penetration involves the fusion of the viral envelope with the plasma membrane of the cell. After fusion, the virus is released into the cytoplasm. They have shown that the fusion process depends on the presence of a specific protease associated with the cell, such as a trypsinlike, elastaselike, or plasminlike protease. If the cell producing the virus lacks the specific protease, then the progeny virus cannot fuse with the next cell, and infection is blocked. This protease-specific fusion mechanism for penetration may be important in the cellular and organ tropism displayed by many enveloped viruses.

45

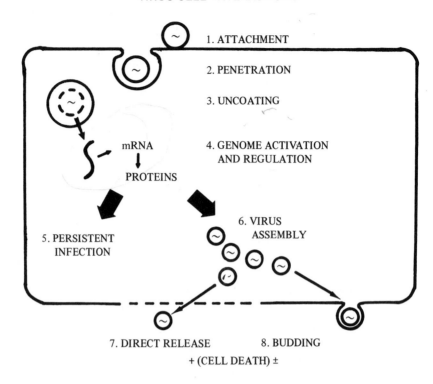

1. ATTACHMENT

2. PENETRATION

3. UNCOATING

mRNA

4. GENOME ACTIVATION
   AND REGULATION

PROTEINS

6. VIRUS
   ASSEMBLY

5. PERSISTENT
   INFECTION

7. DIRECT RELEASE

8. BUDDING

+ (CELL DEATH) ±

Figure 3.1. Phases of virus reproduction cycle. See text for explanation of the eight phases. (Modified from ref. 24.)

3. *Uncoating* is the process whereby the covering of the virus is removed and this step must occur before viral RNA and proteins can be synthesized. Lonberg-Holm and Philipson (1) also reviewed this process in detail. Little is known about the precise mechanisms of uncoating, but cellular enzymes, such as lipases and proteases from lysosomes and/or the cytoplasmic or nuclear sap, are probably important. With the nonenveloped oncogenic SV40, I have demonstrated that the uncoating process requires at least one cellular function as well as ongoing protein synthesis (3,4). Therefore, the uncoating process for this virus is not passive but requires the active participation of the cell. This observation may be true for other viruses as well.

4. *Genome activation and regulation* can occur once the virus is sufficiently uncoated. This phase of infection is responsible for the production of new viral RNA and proteins and has attracted a great deal

46

of investigation (5). The important question for this phase is whether the genome is activated and, if it is, what type and to what extent viral RNA and protein are synthesized.

For many viruses, the pathway may split at this point, depending on the specific virus and the cell with which it is interacting.

5. *Persistent infection* may occur if the production of progeny virus is blocked. In this type of infection, the virus and cell develop a mutually compatible relationship of varying duration. The different types of persistent infection will be discussed below in relation to production of disease.

6. *Virus assembly* occurs when the interaction produces progeny genomes and proteins in an environment that permits the assembly of progeny viruses.

7. *Direct release* is the means by which the new infective viruses reach the extracellular space, killing the cell in some cases.

8. *Budding* from cellular membrane may also occur with or without cell death. The effect of the type of virus release in the production of disease is discussed below.

VIRUS-CELL INTERACTIONS AND THEIR RELATIONSHIP TO DISEASE

Specific types of virus-cell interaction and the clinical diseases produced are shown in Table 3.1. The list is representative but not all-inclusive. Assignment of some of the diseases is somewhat speculative but useful as a basis for discussion. Future elucidation of the molecular mechanisms involved in the pathogenesis of virus-caused disease will clarify and expand the classification.

Type 1. If the virus cannot attach to the plasma membrane because it or the cell lacks the specific receptor, virus infection cannot occur, and no disease will be produced. This is an important mechanism in the cellular and organ tropism of a virus (1).

Type 2. If the virus is able to attach to the cell but cannot penetrate it, no disease is produced. Penetration of the plasma membrane by the virus can fail in the absence of viropexis (pinocytosis of the virus particle) (1) and/or fusion of the viral envelope with the plasma mem-

TABLE 3.1. *Types of Virus-Cell Interaction and Disease Production in Humans*

Virus-Cell Interaction	Mechanism	Disease*
1. Virus cannot attach to cell	1. Lack of virus-specific receptors on plasma membrane	1. None
2. Virus cannot penetrate cell	2A. Pinocytosis (viropexis) not effective B. Membrane fusion not effective (protease-specific)	2. None
3. Virus penetrates cell, but the viral genome is not activated and/or the virus is destroyed	3A. Failure of lysosomal enzymes to activate viral genome B. Failure of cellular function(s) to activate viral genome C. Virion structure prevents proper uncoating and activation	3. None
4. Viral genome is activated, but the virus does not reproduce and the cell is not killed (persistent infection)	4A. Cell not perturbed B. Cellular structure and/or function altered† 1. Altered plasma membrane plus immune attack 2. Immune reaction to virus-specific protein, RNA, or DNA 3. Altered growth control 4. Altered cellular differentiation 5. Altered cellular metabolism	4A. "None"—can get indirect type 5B disease with herpetoviruses B1. SSPE. ?MS, LCM 2. SLE 3. Congenital rubella, warts, NPC, ?PML astrocytes, ?hepatoma, ?leukemias 4. Burkitt lymphoma, ?lymphoma, ?leukemias 5. Subacute spongiform encephalopathies: kuru, CJD, familial Alzheimer dementia; Paget's disease of bone

leukemia

B. Orthomyxoviruses: *Influenza; rhinoviruses, coronaviruses, adenoviruses:* colds; *hepetoviruses;* varicella, encephalitis, cold sores (from type 4A infected sensory neuron); *paramyxovirus:* mumps; *picornaviruses:* polio, myositis, myocarditis, gastroenteritis, hepatitis A; *papovaviruses:* PML oligodendrocytes

6. Measles and hepatitis B: depends on immune status—cytocidal vs. persistent infection; LCM: depends on age

B. Cell killed

6. Mixtures of above types

---

* Examples only, not inclusive. Abbreviations are: CJD—Creutzfeldt-Jakob disease; SSPE—subacute sclerosing panencephalitis; MS—multiple sclerosis; LCM—lymphocytic choriomeningitis; SLE—systemic lupus erythematosus; NPC—nasopharyngeal carcinoma; PML—progressive multifocal leukoencephalopathy.
† Mechanisms 4B1-6 are not mutually exclusive. One mechanism usually predominates in a disease.

brane due to the absence of a cellular protease specific for the infecting virus (2).

Type 3. If the virus penetrates the cell, but the viral genome is not activated or the virus is destroyed or both, no disease is produced. At least three possible mechanisms are responsible for this interaction. (a) If the activation of the virus depends on lysosomal enzymes, then either failure of the virus to combine with a lysosome or failure of one or more lysosomal enzymes to uncoat the virus will result in the functional inactivation of the virus (1). (b) If the activation of the virus does not depend on lysosomal enzymes but does depend on a cytoplasmic or nuclear protein or both for its activation, then either the absence of that protein or its dysfunction will also result in a functional inactivation of the virus (3,4). (c) An altered protein in the virion structure can also inhibit uncoating with the resultant destruction of the virus before genome activation occurs (3,4,6).

Type 4. If the viral genome is activated, but the virus does not reproduce and the cell is not killed (*persistent infection*), the type of disease produced depends on the type of persistent infection that occurs.

4A. If no perturbation of the cell occurs, no disease is produced. Some herpetovirus infections, however, are an exception to this rule (see Chapter 9). Herpes simplex virus can persistently infect neurons in sensory ganglia without perturbing the neuron (type 4A interaction). Occasionally, this "latent" virus undergoes detectable replication without killing the neuron. The newly made virus migrates down the axon to the skin or oral mucous membranes and replicates in and kills the epithelial cells (type 5B interaction), thereby producing cold sores and canker sores. Herpes zoster (shingles) is caused by varicellazoster virus through a similar mechanism. The cytocidal phase in the dermal epithelial cells is controlled by the immune system of the host, virus replication in the sensory neuron is stopped, and the cycle is ready to begin again.

If cellular structure or function or both are altered, then at least six types of interaction can occur with attendant clinical disease.

4B1. The plasma membrane is altered, and the altered cell is acted on by the immune system (7). Three diseases may fall in this category: subacute sclerosing panencephalitis (SSPE), multiple sclerosis, and lymphocytic choriomeningitis (LCM). SSPE is produced by a per-

50

sistent infection of measles virus (8). An *in vitro* measles virus-cell culture model system has been developed by Joseph and Oldstone (9; reviewed by Lampert et al [10]) that may be analogous to the measles virus-cell interaction occurring *in vivo* during SSPE. When human HeLa cells are infected with measles virus and exposed to complement-free measles-specific immunoglobulin G (IgG), the viral antigens in the plasma membrane are redistributed to one end of the cell, forming a "cap." The cap, containing all the viral antigens in the plasma membrane, is shed into the medium, leaving the persistently infected cell with a plasma membrane that has no detectable viral antigens on it. The persistently infected cell is no longer susceptible to immune attack and cytolysis, although the viral genome is replicated and viral antigens are present within the cell. If the measles-specific IgG is removed, viral antigens become detectable on the plasma membrane. Subsequent treatment with complement-free measles-specific IgG again produces capping and shedding of the plasma membrane-associated viral antigens.

In a similar manner, the measles-infected cells in the brain may shed the plasma membrane-associated viral antigens in the presence of measles-specific antibody and become resistant to humoral and cellular immune attack. As the measles-specific antibody level falls after the acute illness, the viral antigens reappear on the plasma membrane, the cell is killed by the immune system, and clinical symptoms are produced. Multiple sclerosis may also be caused in part by a persistent virus infection, possibly by a paramyxovirus (11). A mechanism similar to that described above for SSPE could possibly be involved in the pathogenesis of multiple sclerosis. LCM virus, normally a murine-infecting agent, can infect humans. In mice after the newborn period, the immune response to the infected cell produces the clinical symptoms primarily through an attack on the altered plasma membrane by activated, LCM-specific, T lymphocytes (12).

4B2. If the body's immune response is directed primarily to extracellular viral proteins and nucleic acids and not to the infected cell itself, then antigen-antibody complex (immune complex) disease can occur (13). Systemic lupus erythematosus (SLE) is the best example in this category. This disease may be produced by an immune reaction against the RNA, DNA, and/or proteins of a human C-type retrovirus (14), although independent confirmation is necessary to establish the concept. Whether this human retrovirus is transmitted horizontally as an exogenous virus or is an activated endogenous virus is not known. The antibodies made by the immune system against the retrovirus RNA, DNA, and proteins can produce disease in at least two ways. First, the antibodies directed against the viral antigens may act against similar human antigens, producing the "autoimmune" component of the disease (15,16). Second, the antigen-antibody complexes can be filtered out of the blood by kidney glomeruli, producing the glomerulonephritis characteristic of the disease (13), and by the choroid plexus, possibly producing the transient neurological and mental disturbances known to occur with SLE (17).

4B3. If the virus alters the cell's normal growth-control mechanisms, both benign and malignant disease can be produced. The disease examples for this category are congenital rubella, human warts, and possibly condylomata acuminata, nasopharyngeal carcinoma, the transformed astrocytes found in PML, and hepatoma. Normal cellular division is slowed and finally completely inhibited in human fetal (two- to four-month-old) cells when they are persistently infected with rubella virus (18). This alteration of normal growth control is sufficient to explain the reduction in organ cell number and the production of congenital anomalies in patients with congenital rubella. Human papilloma viruses alter the growth control in epidermal cells, producing common warts and possibly the anogenital or veneral warts called condylomata acuminata. These two types of warts represent benign diseases and are most likely produced by different viruses (19). Nasopharyngeal carcinoma is a malignant disease, possibly produced or promoted by a herpetovirus (Epstein-Barr virus), that alters normal growth control (see Chapter 10). PML is a rare disease produced by the human papovavirus, JCV (see Chapter 11). Although the oligodendrocytes are killed by the replication of the virus, associated astrocytes develop nuclear abnormalities that meet the cytological criteria of premalignant (transformed) nuclei. JCV produces gliomalike cerebral and cerebellar tumors in hamsters (20). BKV, another human papovavirus, has been isolated from a brain tumor in a patient with Wiskott-Aldrich syndrome (21). The pathogenic role of human papovaviruses in human neoplasia has yet to be established, but the above evidence suggests that these viruses do alter normal cellular growth mechanisms and produce human disease. The hepatitis B virus contains a circular double-stranded DNA, a genome similar to the papovavirus genome but much smaller. This virus may be oncogenic in a manner similar to the papovaviruses and may be involved in the pathogenesis of hepatoma (22).

4B4. If the virus alters the normal differentiation of a stem cell, the virus could possibly produce congenital anomalies and neoplastic disease. No evidence exists at present for the role of this type of virus-cell interaction in human disease. Human leukemias and lymphomas may be examples. The role of any virus as a causative agent in these human malignancies has not been demonstrated. EBV has been associated with Burkitt lymphoma, and C-type retroviruses are candidate viruses for human leukemia because they produce these diseases in many other species of animal (see Chapter 12). The possibility exists that leukemia and lymphoma cells are really not "transformed" or structurally altered by the virus but may simply be blocked at an early stage of differentiation, a stage (or stages) associated with a high rate of proliferation.

4B5. If the virus alters the cell's function but not its structure, disease production is possible in organs in which such "dysfunction" of relatively few cells can produce clinical disease (e.g., certain pathways in the brain, osteoclasts in bone). The class of diseases known as subacute spongiform encephalopathies (kuru, Creutzfeldt-Jakob disease, and familial Alzheimer dementia) may be examples (see Chapter 13). There are no detectable inflammatory or immune responses in these diseases. The vacuolation and abnormal membrane components found in affected neurons may cause a metabolic dysfunction and produce clinical disease. The affected neurons are killed, but late in the disease. Paget's disease of bone may also fit into this category. Measles-type nucleocapsids have been observed in the nuclei of osteoclasts in Paget's disease (23). The disease is characterized by increased osteoclastic bone resorption followed by compensatory bone formation. The increased osteoclastic activity may result from a dysfunction of normal cellular metabolism by the persistent measleslike infection.

4B6. Other mechanisms will most likely be found in the future.

Type 5. When the virus reproduces in the cell, the production of disease depends on whether the cell is killed. If the virus simply buds from the plasma membrane without killing the cell (type 5A), a steady-state balance is achieved between the immune attack on the viral antigens and the function of the infected cells. The patient would have a relatively normal life until immune complex disease occurred. No human disease can yet be placed in this category. LCM of mice does fit in this category (12), and this virus can infect humans. Many non-human leukemias produced by C-type retroviruses are included in this category because the leukemic cells continue to produce virus and divide throughout the illness.

Many of the classic virus-caused human diseases result when the virus reproduces and kills the cell (type 5B). Some examples follow: ortho-myxoviruses—influenza; paramyxoviruses—mumps; rhinoviruses, coro-naviruses, adenoviruses—colds; herpesviruses—varicella-zoster, cold sores, canker sores, encephalitis; picornaviruses—poliomyelitis, myo-carditis, nasopharyngitis, gastroenteritis, myositis, hepatitis A; papo-vaviruses—the oligodendrocyte in PML demyelination.

Type 6. Mixtures of the above virus-cell interactions can produce disease, although one of the above interactions usually predominates. One example of human disease is the production of acute measles (type 5B) in contrast to the production of subacute sclerosing pan-encephalitis (type 4B1). A second example of human disease in this category is viral hepatitis B. The course of disease depends on whether the acute virus-caused hepatocyte death (type 5B) is survived and the virus eradicated or whether a persistent infection is established with a subsequent variable reaction of the immune system to the persistently infected cells (types 4B1, 2, 3, and 5). LCM virus produces acute pan-encephalitis in adult mice (type 5B) and a persistent infection in new-born mice that is not cytocidal (type 5A). The neonatal disease does produce immune complex disease (type 4B2), however.

A RESEARCH MODEL: THE MURINE HEPATOENCEPHALITIS VIRUS SYSTEM

Murine coronaviruses (murine hepatitis) are being used in my labora-tory as a model to identify the precise molecular mechanisms of virus reproduction, virus protein and RNA structure, and virus-cell inter-action that may be involved in the pathogenesis of different disease states. The members of the murine hepatoencephalitis coronavirus group were originally referred to as mouse hepatitis virus (MHV). Each strain of virus causes a more or less specific spectrum of disease. Some strains produce only hepatitis with or without cirrhosis, some encephalitis with or without demyelination, some interstitial pneumonia, and some are not pathogenic. The pathogenesis of a specific disease in this spectrum depends also on the strain and age of the mouse and the route of infection. For example, one unit of infective virus is sufficient to produce lethal panencephalitis when given to four-week-old BALB/c mice by the intracerebral route.

Naturally occurring variants or mutants of MHV are listed in Table 3.2. Thirty-four experimentally produced temperature-sensitive mutants have been isolated in my laboratory from the MHV-4 (JHM strain). These mutants are inhibited at various stages of reproduction *in vitro* at 38.5°C but are completely normal when grown at 33°C. These mutants produce modified disease in mice, in which a body temperature of 37° to 38°C is inhibitory for the mutants. The naturally occurring and experimentally produced mutants can be used to study the pathogenesis of the various disease states at cellular and molecular levels using intranasal, intracerebral, and intraperitoneal infections. The phases of the viral reproduction cycle (Fig. 3.1), the virus-specific proteins and RNA, and the interactions of the mutants with cells *in vitro* at 33° and 38.5°C and in mice of various ages can thus be analyzed.

The molecular mechanisms elucidated by this approach should be applicable to virus-induced human disease because the human and murine coronaviruses are antigenically related and probably produce similar disease in their respective hosts. Human coronaviruses are the second most common cause of colds, just after rhinoviruses. The human

TABLE 3.2. *Organ Tropism of Murine Hepatoencephalitis Coronaviruses*

Virus	Liver (Hepatitis)	Brain (Encephalitis)	Lung (Pneumonia)
MHV-1*	+	0	0
MHV-2	+ (cirrhosis)	0	0
MHV-3	+ (no cirrhosis)	+ (IP)	+ (IP)
MHV-4 (JHM)	0	+ (IN)	+ (IN)
MHV-A59	+	0	0
MHV-A59 (avirulent)	0	0	0

* MHV—mouse hepatitis virus. Other members of this group include: MHV-B, MHV (BALB/c), MHV-S, MHV (SR1), MHV (SR2), MHV (SR3), and MHV (SR4). IP—interperitoneal infection, IN—intranasal infection. Data obtained from Piazza M: *Experimental Viral Hepatitis,* Springfield, Ill., Charles C Thomas, 1969.

coronaviruses may also produce interstitial pneumonia, gastroenteritis, and hepatitis in humans because coronaviruses of other species produce similar diseases in their respective hosts.

CONCLUSION

The goal of this classification has been to provide a beginning structure with which to understand the molecular mechanisms involved in the pathogenesis of virus-caused human disease. The viral replication cycle provides a logical framework for the description of the pathogenesis of virus-caused human disease at the molecular level. Correlations between the biochemistry and the biology of various virus-cell and virus-animal systems will elucidate the molecular mechanisms. As investigations progress, new viruses will be discovered, and more than one currently categorized "idiopathic" human disease will be found to be the result of these virus-cell interactions.

REFERENCES

1.   Lonberg-Holm K, Philipson L: Early interaction between animal viruses and cells. *Monogr Virol* vol 9, 1974.
2.   Scheid A, Choppin PW: Protease activation mutants of Sendai virus: Activation of biological properties by specific proteases. *Virology* 69:265–277, 1976.
3.   Robb JA, Martin RG: Genetic analysis of simian virus 40: III. Characterization of a temperature-sensitive mutant blocked at an early stage of productive infection in monkey cells. *J Virol* 9:956–968, 1972.
4.   Robb JA, Huebner K: Effect of cell chromosome number on simian virus 40 replication. *Exp Cell Res* 81:120–126, 1973.
5.   Fenner F, McAuslan BR, Mims CA, et al: *The Biology of Animal Viruses,* ed 2. New York, Academic Press, 1974.
6.   Robb JA, Lopez-Revilla R: Temperature-dependent synthesis of early and late SV40 cytoplasmic RNA during *tsD*101* mutant virion infection. *Intervirology* 6:122–128, 1976.
7.   Notkins AL: *Viral Immunology and Immunopathology.* New York, Academic Press, 1975.
8.   ter Muelen V, Katz M, Muller D: Subacate sclerosing panencephalitis: A review. *Curr Topics Microbiol Immunol* 57:1–38, 1972.
9.   Joseph BS, Oldstone MBA: Immunologic injury in measles virus infection: II. Suppression of immune injury through antigenic modulation. *J Exp Med* 142:864–867, 1975.
10.   Lampert PW, Joseph BS, Oldstone MBA: Morphological changes of

cells infected with measles or related viruses, in Zimmerman H (ed): *Progress in Neuropathology*. New York, Grune and Stratton, 1975, vol 3, pp 51–68.

11.   Johnson RT: Virological data supporting the viral hypothesis in multiple sclerosis, in *Medical Research Council-Multiple Sclerosis Research*. London, Her Majesty's Stationery Office, 1975, pp 155–183.

12.   Holland JJ: Slow, inapparent and recurrent viruses. *Sci Am* 230:33–40, 1974.

13.   Oldstone MBA, Dixon FJ: Immune complex disease associated with viral infections, in Notkins AL (ed): *Viral Immunology and Immunopathology*. New York, Academic Press, 1975, pp 341–356.

14.   Panem S, Ordonez NG, Kirsten WH, et al: C-type virus expression in systemic lupus erythematosus. *N Engl J Med* 295:470–475, 1976.

15.   Marx JL: Autoimmune disease: New evidence about lupus. *Science* 192:1089–1091, 1976.

16.   Schwartz RS: Viruses and systemic lupus erythematosus. *N Engl J Med* 293:132–136, 1975.

17.   Lampert PW, Oldstone MBA: Pathology of the choroid plexus in spontaneous immune complex disease and chronic viral infections. *Virchows Arch [Pathol Anat]* 363:21–32, 1974.

18.   Boué A, Plotkin SA, Boué JG: Action du virus de la rubéole sur différénts systémes de cultures de cellules embryonnaires humaines. *Arch Gesamte Virusforsch* 16:443–458, 1965.

19.   Delap R, Friedman-Kien A, Rush MG: The absence of human papilloma viral DNA sequences in condylomata acuminata. *Virology* 74:268–272, 1976.

20.   Walker DL, Padgett BL, ZuRhein Gm, et al: Human papovavirus (JC): Induction of brain tumors in hamsters. *Science* 181:674–676, 1973.

21.   Takemoto KK, Rabson AS, Mullarkey MF, et al: Isolation of papovavirus from brain tumor and urine of a patient with Wiskott-Aldrich syndrome. *J Natl Cancer Inst* 53:1205–1207, 1974.

22.   Jawetz E, Melnick JL, Adelberg EA: *Review of Medical Microbiology*, ed 12. Los Altos, Calif, Lange Med Publ, 1976, p 490.

23.   Mills BG, Singer FR: Nuclear inclusions in Paget's disease of bone. *Science* 194:201–202, 1976.

24.   Fenner FJ, White DO: *Medical Virology*. New York, Academic Press, 1976.

4

Viral Hepatitis

SAUL KRUGMAN

When the term "viral hepatitis" was first used in the 1940s, it referred to two distinct diseases: "infectious hepatitis" and "serum hepatitis." In 1947, MacCallum in Great Britain proposed the designation of type A hepatitis for infectious hepatitis and type B hepatitis for serum hepatitis. By the end of the 1960s, this nomenclature was adopted when two etiologically and immunologically distinct agents were identified as causes of these diseases: hepatitis A virus (HAV) and hepatitis B virus (HBV). More recently, it has become clear that hepatitis can be caused by a viruslike agent or agents proved to be neither HAV nor HBV. Those agents have not been characterized as yet.

The studies by various investigators in the 1940s and early 1950s revealed that type A hepatitis had a relatively short incubation period of 15 to 40 days, it was usually transmitted from person to person via the fecal-oral route, and occasionally it was spread by the parenteral route or by ingestion of contaminated food or water (1–7). In contrast, type B hepatitis appeared to spread exclusively via the parenteral route and presumably did not spread from person to person by contact.

During the 1960s, we identified two immunologically distinct types of hepatitis that were designated MS-1 and MS-2 (8). It soon became apparent that MS-1 was typical type A hepatitis and MS-2 was probably type B hepatitis. However, unlike the finding in earlier studies (1–3), the MS-2 strain of HBV was proved to be infectious by mouth. Additional epidemiological studies showed that type B hepatitis, like type A, spread from person to person by contact.

During the past decade, extraordinary advances have occurred in hepatitis research, advances that have further clarified the natural history of type A and type B hepatitis.

TYPE A HEPATITIS: RECENT DEVELOPMENTS

During the past 25 years, many virologists have tried and failed to cultivate human HAV in cell or organ cultures. In addition, until the mid 1960s, transmitting the virus to animals seemed impossible. However, during the past 10 years, HAV has been successfully transmitted to such nonhuman primates as marmoset monkeys and chimpanzees (9–13).

Hepatitis A virus was first visualized by immune electronmicroscopy in 1973 by Feinstone and colleagues (14), who examined stool filtrates from patients with type A hepatitis. They identified 27-nm viruslike particles that were shown to be immunologically related to HAV.

Inoculation of stool filtrates containing these 27-nm particles induced type A hepatitis in susceptible marmoset monkeys and chimpanzees (15). HAV was further characterized by Hilleman and colleagues (16), who found that the agent was probably an RNA virus that had the physical, chemical, and biologic characteristics of an enterovirus. An electronmicrograph of HAV as compared with HBV is shown in Figure 4.1.

The availability of hepatitis A antigen, derived from marmoset-infected liver, enabled Hilleman's group to develop a specific immune adherence hemagglutination test for the detection of hepatitis A antibody (17). This specific test enabled our group to determine the time of appearance and persistence of hepatitis A antibody by examining serial serum specimens obtained from 20 patients having type A hepatitis (18). The results in one patient, shown in Figure 4.2, are representative of the entire group. Hepatitis A antibody appeared one to four weeks after onset of infection, and it persisted for many years. As

Figure 4.1.   Electron micrograph of type A and type B hepatitis viruses. Type A: note 27-nm particles, uniform in size. Type B: note 42-nm Dane particles (hepatitis B virus) and spherical and filamentous particles 20 nm in diameter (hepatitis B surface antigen). (From ref. 16.)

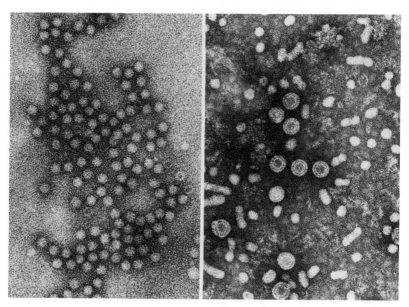

VIRAL HEPATITIS TYPE A

IMMUNE ADHERENCE ANTIBODY RESPONSE

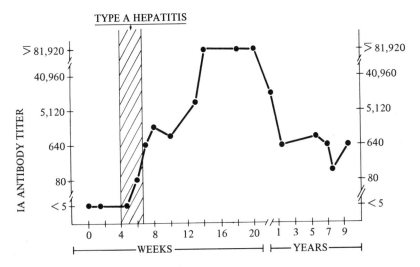

Figure 4.2. Viral hepatitis, type A: results of serial tests for hepatitis A antibody by a specific immune adherence test. (From ref. 18.)

indicated in Figure 4.3, abnormal serum glutamic oxaloacetic transaminase (SGOT) values were first detected 31 days after exposure, and hepatitis A antibody was detected 12 days later on day 43.

More recently, a radioimmunoassay procedure has been developed for the detection of hepatitis A antibody (19). When adequate supplies of antigen and other reagents become available, it will be possible (a) to confirm a diagnosis of type A hepatitis, (b) to determine whether a person is susceptible or immune, and (c) to measure the quantity of hepatitis A antibody in gamma globulin preparations.

Prevention of Type A Hepatitis. Passive immunization procedures for the prevention of type A hepatitis have been available for about 30 years. Standard preparations of human immune serum globulin (ISG) have been effective in the prevention or modification of the infection. A hepatitis A vaccine for active immunization has not been developed thus far because HAV has not been successfully cultivated in cell culture.

HEPATITIS A VIRUS INFECTION

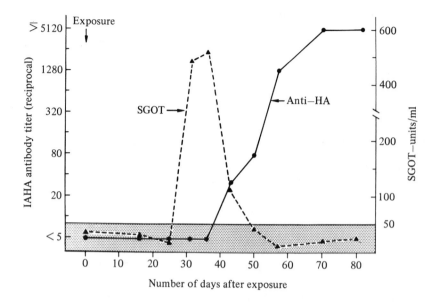

Figure 4.3. Appearance of abnormal SGOT values on day 31 and first detection of immune adherence hepatitis A antibody (anti-HAV) on day 43 of typical anicteric HAV infection. (From Krugman S: *J Infect Dis* 134:70, 1976.)

TYPE B HEPATITIS: RECENT DEVELOPMENTS

The discovery of Australia antigen by Blumberg et al (20), its subsequent association with HBV by others (21,22), and important contributions by many other investigators provided the technology that led to (a) the identification and characterization of HBV, (b) the development of highly sensitive tests to detect various hepatitis B antigens and antibodies, (c) the selection of susceptible chimpanzees to serve as animal models, (d) the preparation of an inactivated hepatitis B vaccine for the prevention of the disease, and (e) the evaluation of interferon for the treatment of chronic hepatitis B infection.

Hepatitis B Virus and Antigens. Electronmicroscopic examination of serum from a patient with type B hepatitis reveals the following types of viruslike particles: (a) spherical particles, 20 nm in diameter, (b) filamentous particles, 100 nm long or longer and 20 nm in diameter, and (c) 42-nm Dane particles (23–25) (see Fig. 4.1B). Impressive indirect evidence indicates that the Dane particle is the complete hepatitis B virion, and the smaller 20-nm particles may represent excess virus-coat material.

At least three immunologically distinct antigens are associated with the Dane particle: (a) the hepatitis B surface antigen (HBsAg), formerly called Australia antigen and hepatitis-associated antigen (HAA); (b) the hepatitis B core antigen (HBcAg); and (c) the e antigen (HBeAg). These antigens induce the production of their respective antibodies, anti-HBs, anti-HBc, and anti-HBe. The HBsAg is located in the surface component of the Dane particle. The core component contains HBcAg, HBeAg, DNA polymerase, and double-stranded DNA.

The use of various tests to detect hepatitis B antigens and antibodies has added considerably to knowledge of the natural history of this disease. It has been possible to chart the course and consequences of hepatitis B infection from the time of exposure to onset of infection and for many months and years thereafter. Most patients with type B hepatitis recover completely. Occasionally, however, the disease may progress to either chronic hepatitis or an asymptomatic chronic carrier state. In rare instances, a rapid, fulminating, fatal outcome may occur.

Acute Type B Hepatitis with Recovery. Figure 4.4 shows the course of a patient with acute type B hepatitis. The first indication of infection is the appearance of HBsAg in the blood, usually noted about four weeks after exposure. About one week later, HBeAg and DNA polymerase activity are detectable. Evidence of liver dysfunction (abnormal SGOT) is observed about two months after exposure, or about one month after HBsAg becomes detectable. Jaundice, when present, appears two to four weeks after the SGOT levels become abnormal.

The first antibody to be detected is anti-HBc; it appears about the time that jaundice is noted. The protective antibody, anti-HBs, appears several weeks to several months after HBsAg is no longer detectable. In most persons, anti-HBs persists for many years, but occasionally it declines to undetectable levels. In addition, about 20% of persons with proved type B hepatitis do not develop detectable anti-HB$_s$. This phenomenon may be a reflection of the sensitivity of the test.

VIRAL HEPATITIS – TYPE B

ACUTE INFECTION FOLLOWED BY RECOVERY

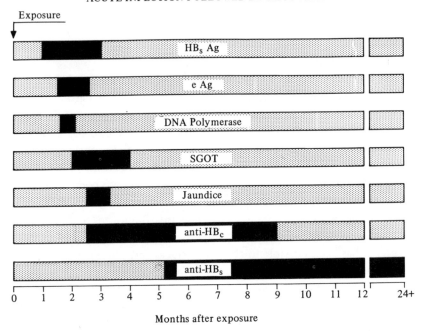

Figure 4.4.   Acute viral hepatitis, type B, followed by recovery: results of serial tests for HBsAg, HBeAg, DNA polymerase, SGOT, jaundice, anti-HBc, and anti-HBs. Black areas = detectable or abnormal; stippled areas = not detectable or normal. (From Krugman, Ward, and Katz: *Infectious Diseases of Children,* ed 6. St. Louis, C V Mosby Co, 1977.)

Acute Type B Hepatitis Followed by Chronic Active Hepatitis. Figure 4.5 is a schematic illustration of the course of type B hepatitis that progresses to chronic active hepatitis. In this example the HBsAg, HBeAg, DNA polymerase activity, abnormal SGOT values, and anti-HBc persisted for more than two years. The detection of various antigens and antibodies depends on the sensitivity of the test. For example, the immunodiffusion test used to detect HBeAg is not very sensitive. Therefore, certain patients having HBsAg-positive chronic active hepatitis may not have detectable HBeAg.

Acute Type B Hepatitis Followed by HBsAg Carrier State. Figure 4.6 is a schematic illustration of the course of a patient whose acute hepa-

VIRAL HEPATITIS – TYPE B

ACUTE INFECTION FOLLOWED BY CHRONIC ACTIVE HEPATITIS

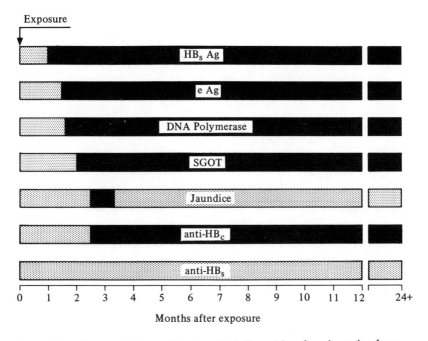

Figure 4.5. Acute viral hepatitis, type B, followed by chronic active hepatitis: results of serial tests for HBsAg, HBeAg, DNA polymerase, SGOT, jaundice, anti-HBc, and anti-HBs. (See Fig. 4.4 for key.)

titis was followed by an asymptomatic HBsAg-positive carrier state. The HBeAg, DNA polymerase, and abnormal SGOT levels are transient. Later, anti-HBc and anti-HBe levels appear and persist. In spite of the persistence of the HBsAg carrier state, follow-up biopsies usually show complete resolution of the pathological findings that were present during the acute stage of the disease.

Non-A, Non-B Acute Viral Hepatitis. Figure 4.7 is a schematic illustration of the course of a patient whose post-transfusion hepatitis was shown to be neither type A nor type B. Hepatitis with jaundice occurred 60 to 75 days after exposure to a blood transfusion. Results of tests for HBsAg, anti-HBc, anti-HBs, and anti-HAV were negative. In addition, patients with non-A, non-B hepatitis have been tested for the

VIRAL HEPATITIS – TYPE B

ACUTE INFECTION FOLLOWED BY HBS Ag CARRIER STATE

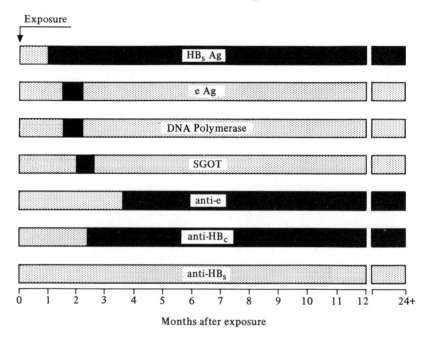

Figure 4.6. Acute viral hepatitis, type B, followed by asymptomatic HBs-Ag carrier state: results of serial tests for HBsAg, HBeAg, DNA polymerase, SGOT, anti-e, anti-HBc, and anti-HBs. (See Fig. 4.4 for key.)

presence of cytomegalovirus and Epstein-Barr virus infection to rule out these viral causes of hepatitis.

Extrahepatitic Manifestations of Type B Hepatitis. Type B hepatitis infection has been shown to be associated with these syndromes: (a) a serum-sicknesslike prodrome, (b) polyarteritis nodosa, and (c) glomerulonephritis (26).

The serum-sickness prodrome is characterized by a transient, erythematous, maculopapular eruption with or without urticaria, polyarthralgia, and occasionally actual arthritis. These manifestations usually occur several days or weeks before the onset of acute hepatitis. A transient suppression of complement titer and of C3 and C4 may be noted during the early phase of the skin and joint manifestations.

68

NON–A NON–B ACUTE VIRAL HEPATITIS

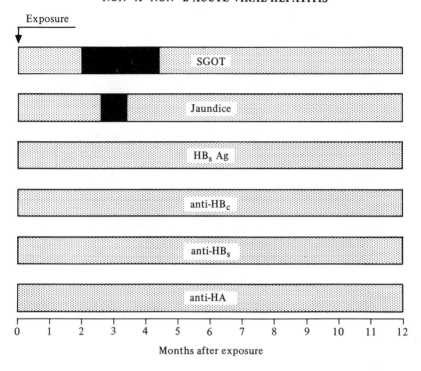

Figure 4.7. Non-A, non-B acute viral hepatitis: results of serial tests for SGOT, jaundice, HBsAg, anti-HBc, anti-HBs, and anti-HA. (See Fig. 4.4 for key.)

Polyarteritis nodosa associated with type B hepatitis is characterized by fever, polyarthralgia, myalgia, rash, and urticaria. The syndrome, including peripheral neuropathies, hypertension, and renal lesions, may evolve during a period of months. The observation of a HBsAg-carrier state in these patients was described by Gocke (26) and Trepo and Thivolet (27).

Various investigators have demonstrated an association of glomerulonephritis with type B hepatitis (28,29). Nodular deposits of HBsAg immunoglobulin and C3 in the glomeruli have been seen by electron-microscopy. The glomerulonephritis is usually of the membranous or membranoproliferative type. Most patients have chronic active hepatitis and hepatitis B antigenemia.

PREVENTION OF TYPE B HEPATITIS

Passive Immunization. The efficacy of standard ISG for the prevention or modification of type B hepatitis has been a controversial subject for many years. Various studies, chiefly in patients who received multiple transfusions, indicated that ISG provided little or no protection. In retrospect, however, it is now known that the ISG used for these studies had little or no detectable anti-HBs. Moreover, it is also known that other viruses can cause post-transfusion hepatitis.

Since 1972, most lots of standard ISG have had higher levels of antibody, and a special high-titer hepatitis B ISG (HBIG) has been developed. Recent studies have shown that HBIG may be effective for the prevention or modification of type B hepatitis (30–33). When this preparation is licensed for use, the recommended dosage will depend on the antibody titer of the preparation.

Active Immunization. The development of an inactivated hepatitis B vaccine occurred in spite of the failure to cultivate HBV in cell culture. Our studies with heat-inactivated MS-2 serum containing HBV showed that it was possible to destroy the infectivity of a solution of MS-2 serum in water without affecting its antigenicity (34). Additional studies showed that active immunization with the heat-inactivated MS-2 serum was associated with induction of antibody (anti-HBs) and a protective effect (35,36). It appeared, therefore, that it should be feasible to develop a vaccine from purified preparations of HBsAg.

The successful transmission of HBV to chimpanzees provided an animal model for the evaluation of the safety, antigenicity, and efficacy of experimental hepatitis B vaccines (37). The results of hepatitis B vaccine studies in chimpanzees have been described by Purcell and Gerin (38), Hilleman et al. (39), and Buynak et al. (40). Those studies showed that formalin-inactivated hepatitis B vaccine prepared from purified HBsAg was not infectious, but it was antigenic and protective for chimpanzees.

During the past year, in collaboration with Hilleman's group, we administered the formalin-inactivated hepatitis B vaccine to 11 seronegative adults. Serial observations for a 10-month period showed no detectable HBsAg or anti-HBc and no evidence of liver dysfunction.

70

This Phase I study confirmed the safety of the vaccine with regard to its lack of infectivity.

Maupas et al. (41), in France, reported the results of studies in patients and staff members of a renal dialysis unit. That preliminary report indicated that the vaccine was not infectious and was antigenic. The studies are still in progress.

CHEMOTHERAPY FOR CHRONIC ACTIVE HEPATITIS

Two recent reports have shown that parenteral administration of human leukocyte interferon for the treatment of chronic hepatitis B infection and chronic active hepatitis has been associated with a suppression of hepatitis B core antigen in serum and in liver (42,43). The effect was transient, but it appeared to be more lasting when interferon was given for a prolonged time. If these preliminary observations are confirmed, the use of interferon may prove to be of value in limiting infectivity or eradicating chronic infection.

SUMMARY AND CONCLUSIONS

During the past decade, new developments in hepatitis research have shed new light on the natural history of the disease and enhanced prospects for prevention by active and passive immunization. Recent advances in HAV research include (a) identification of the causative agent as a 27-nm particle that appears to be a RNA virus with characteristics resembling an enterovirus; (b) successful transmission of HAV to marmoset monkeys and chimpanzees; and (c) development of specific immune adherence hemagglutination and radioimmunoassay antibody tests for identification of hepatitis A infection and for seroepidemiological surveys to determine susceptibility or immunity.

Recent advances in hepatitis B research include (a) identification of the causative agent as a 42-nm particle (Dane particle) that contains (i) an outer coat, the hepatitis B surface antigen; (ii) an inner core, the hepatitis B core antigen; and (iii) DNA polymerase and double-stranded DNA, and hepatitis B e antigen; (b) development of specific, sensitive tests to detect hepatitis B antigens and their respective antibodies; (c) detection of HBsAg and HBeAg as markers of infectivity;

71

(d) transmission of HBV to chimpanzees; (e) transmission of HBV infection from mother to infant in the prenatal, perinatal, and postnatal periods; (f) development of a specific hepatitis B immune serum globulin; and (g) development of an inactivated hepatitis B vaccine.

The availability of specific tests to identify HAV and HBV as causative agents of viral hepatitis has revealed the presence of "non-A, non-B" hepatitis. The agent or agents responsible for this infection have not yet been characterized.

A special hepatitis B immune serum globulin containing high titers of hepatitis B antibody will soon be licensed for use in the United States. The gamma globulin preparation will be indicated for (a) persons who are accidentally inoculated with blood or blood products known to be contaminated with HBV and (b) newborn infants whose mothers either are hepatitis B carriers or had type B hepatitis during the last trimester of pregnancy.

An inactivated hepatitis B vaccine has been shown to be safe (not infectious), antigenic, and protective for chimpanzees. The initial Phase I trial in 11 susceptible adults showed that the vaccine was well tolerated (no local or systemic reactions) and safe (noninfectious). A recent report from France confirmed the safety and antigenicity of the vaccine in patients and staff members of a renal dialysis unit. The initial studies with hepatitis B vaccine are encouraging.

Studies currently in progress indicate that the use of human leukocyte interferon may prove to be of value for the treatment of chronic hepatitis B infection. This chemotherapeutic agent is being evaluated for its efficacy in limiting infection and eradicating chronic infection.

DISCUSSION

Question: A small but definite number of cases of hepatitis are complicated by immune complex formation manifested as vasculitis. Does the consumption of antigens such as HBsAg antigen and antibody directed against it interfere with the diagnosis or monitoring of hepatitis by the serological methods you describe?

Dr. Krugman: If you refer to the diagnosis of hepatitis B infection, I think the crucial factor is to have the appropriate specimens available. If you see the patient early, if you have a sample of blood at that time, and if you can obtain follow-up specimens, a specific diagnosis should not be difficult, because even if immune complexes are present, HBsAg is always, or nearly always, detectable during the latter part of the incubation period. Occasionally, HBsAg may no longer be detectable within a matter of one or two days after onset of jaundice. As a matter of fact, on rare occasions we have seen antigens disappear before jaundice was detectable. These are the exceptions. In most patients, the antigen persists, and, if the reagents become available to test for not only anti-HBsAg but also anti-HBc, diagnosing hepatitis B infections should be reasonably simple.

Question: Dr. Krugman stated that non-A, non-B viral hepatitis should not be called type C. The question is whether it should be called viral hepatitis at all. What is the evidence that the pathogen is a virus, and if it is a virus but not type A or B and not cytomegalovirus, then why not call it type C?

Dr. Krugman: One reason the disease has been called non-A, non-B viral hepatitis is its unusual incubation period. Such prolonged incubation is the kind of circumstantial evidence we had years ago with type A and type B hepatitis. In any event, post-transfusion non-A, non-B hepatitis has been observed after an incubation period of about two months. Neither bacteria nor other viral agents such as cytomegalovirus or EBV has been detected. On the other hand, Dr. Melnick is absolutely right; no specific evidence exists that the disease is truly viral hepatitis, although it seems to have the epidemiological footprints and markers of viral hepatitis. No drug has been implicated in these instances, and no evidence exists that it is a toxic hepatitis. The diagnosis of non-A, non-B hepatitis has been made by exclusion.

Dr. Melnick: It's unwieldy to call it non-A and non-B. If you've got to call it viral hepatitis, I would call it type C until proved otherwise.

Question: What precautions can be taken among health-care professionals, such as dentists, who are healthy carriers?

Dr. Krugman: Singling out dentists is unfair, because surgeons and other health personnel may also be healthy carriers. It should be emphasized that the only way a person can transmit hepatitis is by actual inoculation of blood, body secretions, or other material that contain the virus. A dentist or surgeon who has good technique and can avoid transferring his blood—and the situation would be one where he accidentally cut himself—would really involve no risk. The Public Health

Service Advisory Committee or Immunization Practices and the Committee on Viral Hepatitis of the National Academy of Sciences met about a year ago to review this problem. A supplement to the *Weekly Morbidity and Mortality Report* deals with this problem in great detail, highlighting the various precautions that should be taken. I think the message is that persons who are carriers should not be considered to be lepers, provided that the appropriate precautions can be taken where they are not a menace to their patients and colleagues.

On the other hand, if it becomes obvious that a particular oral surgeon or other health professional cannot or will not take the appropriate precautions, then something must be done to deal with this difficult problem. The precaution of double gloving is appropriate for minimizing the possibility of spreading the infection. I doubt that droplets are a serious threat. The carrier would actually have to expectorate large quantities of material that would then need to be ingested by the recipient. Hepatitis B virus does not fly through the air with the greatest of ease. Unlike the influenza virus, its transmission requires very close physical and intimate contact. Such transmission has been observed in a family in which one spouse has hepatitis.

Dr. Melnick: Another test can be done on carriers, the HBeAg test. If those who are known to be carriers of the HBsAg antigen can be examined for e antigen and it is missing, they are less likely to be infectious than those who carry both HBsAg and HBeAg.

Question: Would you comment on the reverse situation, the surveillance of patients having hepatitis as a risk to health professionals.

Dr. Krugman: I think authorities agree on various points. For example, surveillance should be routine in certain areas where patients are monitored. For patients on renal dialysis units, one ought to attempt to control hepatitis; you have to know what is going on, not only with your patients but also with your staff.

We have the technology available to detect not only antigen but also antibody, so that we can actually identify those persons working on the unit who may have had clinical or subclinical hepatitis in the past and who are immune and thus not likely to get the disease. For certain patients and staff, however, careful monitoring becomes necessary, because specific additional precautions may be taken if carriers are identified.

74

The glove that is supposed to protect both patient and surgeon may be punctured by a needle, or some other accident may occur. A person who happens to be both a carrier and a surgeon has the responsibility for taking additional precautions.

Question: What special concern should be shown for the patient who has hepatitis during pregnancy?

Dr. Krugman: Many investigators have shown that, given two situations, either the occurrence of type B hepatitis during the last trimester of pregnancy or the occurrence of the carrier state all through pregnancy, there is a risk either to the fetus or to the newborn infant. In the United States, the highest risk has been associated with hepatitis during the last trimester of pregnancy, and some of the reports suggest that the likelihood of hepatitis occurring in the infant is at least 50% or more. Initially, in this country, the likelihood of hepatitis occurring in the newborns of mothers who were just carriers and did not have the acute disease was reported to be about 10% or 15%. Most of these studies come from Dr. Schweitzer and his colleagues in California.

In other countries of the world (such as Taiwan and also, I believe, Japan), where the carrier rate is high even in asymptomatic mothers who are carriers, the likelihood of hepatitis occurring in the newborn is about 50%.

What happens to the newborn? The possibilities are several. Nothing may happen, despite exposure, and the infant may go right through the first year of life with no evidence of hepatitis. Or the infant may become infected, which happens in many developing areas of the world, and within about four weeks to three or four months may become antigen positive and remain so. As a matter of fact, there is speculation that these are the infants in areas in Africa and other parts of the world who subsequently develop hepatoma. It has been speculated that the high carrier rate in certain parts of the world where hepatoma is common may be associated with this particular carrier rate, because the correlation between hepatitis B infection and hepatoma is strong. Furthermore, an infant may develop chronic, active, or fulminant hepatitis. Also, successive deaths have occurred in infants born of the same mother. In one mother, there had been four successive pregnancies; the first three children had histories of hepatitis in infancy and died, and the fourth one subsequently developed hepatitis. By use of the technology available, it became obvious that the mother was a carrier.

Dr. Gellis and his group had a report of the same mother giving birth to two infants, both having hepatitis. Several similar occurrences have been reported in the New York area, and this is a real problem.

What can be done about this problem and how can it be prevented? If hepatitis B immune serum globulin is available, it should be given to the infant shortly after birth. If HBIG is not available, standard immune serum globulin should be used. Current lots of standard ISG have more antibody than they had formerly. Studies on this problem are in progress, for example, that of Dr. Szmuness and his colleagues in Taiwan. If the mother is known to be a carrier, it is important to monitor her infant at birth. It is also important to emphasize that nearly all these infants are normal when they are born, because hepatitis occurs after a variable incubation period, depending on whether the infants have been infected *in utero* or at the time of birth. As a matter of fact, most infants for some peculiar reason, are infected at birth—whether they ingest infectious material or are inoculated with the blood—and they do not develop hepatitis for some months later.

REFERENCES

1. MacCallum FO, Bradley WH: Transmission of infective hepatitis to human volunteers. *Lancet* 2:228, 1944.
2. Havens WP Jr, Ward R, Drill VA, et al: Experimental production of hepatitis by feeding icterogenic materials. *Proc Soc Exp Biol Med* 57:206–208, 1944.
3. Paul JR, Havens WP Jr, Sabin AB, et al: Transmission experiments in serum jaundice and infectious hepatitis. *JAMA* 128:911–915, 1945.
4. Stokes J Jr, Berk JE, Malamut LL: The carrier state in viral hepatitis. *JAMA* 154:1059–1066, 1954.
5. Murray R: Viral hepatitis. *Bull NY Acad Med* 31:341–358, 1955.
6. Ward R, Krugman S, Giles JP, et al: Infectious hepatitis: Studies of its natural history and prevention. *N Engl J Med* 258:407–416, 1958.
7. Krugman S, Ward R, Giles JP, et al: Infectious hepatitis: Detection of virus during the incubation period and in clinically inapparent infection. *N Engl J Med* 261:729–734, 1959.
8. Krugman S, Giles JP, Hammond J: Infectious hepatitis: Evidence for two distinctive clinical, epidemiological, and immunological types of infection. *JAMA* 200:365–373, 1967.
9. Deinhardt F, Holmes AW, Capps RB, et al: Studies on the transmission of human viral hepatitis to marmoset monkeys: I. Transmission of disease, serial passages, and description of liver lesions. *J Exp Med* 125:673–687, 1966.
10. Holmes AW, Wolfe L, Rosenblate H, et al: Hepatitis in marmosets:

Induction of disease with coded specimens from a human volunteer study. *Science* 165:816–817, 1969.

11. Lorenz D, Barker L, Stevens D, et al: Hepatitis in the marmoset, Saguinus mystax. *Proc Soc Exp Biol Med* 135:348–354, 1970.

12. Mascoli CC, Ittensohn OL, Villarejos VM, et al: Recovery of hepatitis agents in the marmoset from human cases occurring in Costa Rica. *Proc Soc Exp Biol Med* 142:276–282, 1973.

13. Provost, PF, Ittensohn OL, Villarejos VM, et al: Etiologic relationship of marmoset-propagated CR 326 hepatitis A virus to hepatitis in man (37220). *Proc Soc Exp Biol Med* 142:1257–1267, 1973.

14. Feinstone SM, Kapikian AZ, Purcell RH: Hepatitis A: Detection by immune electron microscopy of a virus-like antigen associated with acute illness. *Science* 182:1026–1028, 1973.

15. Maynard JE, Lorenz D, Bradley DW, et al: Review of infectivity studies in non human primates with virus-like particles associated with MS-1 hepatitis. *Am J Med Sci* 270:81–85, 1975.

16. Provost PF, Wolanski BS, Miller WJ, et al: Biophysical and biochemical properties of CR 326 human hepatitis A virus. *Am J Med Sci* 270:87–92, 1975.

17. Miller WJ, Provost PJ, McAleer WJ, et al: Specific immune adherence assay for human hepatitis A antibody: Application to diagnostic and epidemiologic investigations. *Proc Soc Exp Biol Med* 149:254–261, 1975.

18. Krugman S, Friedman H, Lattimer C: Viral hepatitis, type A: Identification by specific complement fixation and immune adherence tests. *N Engl J Med* 292:1141–1143, 1975.

19. Dienstag JL, Krugman S, Wong DC, et al: Comparison of serological tests for antibody to hepatitis A antigen, using coded specimens from individuals infected with the MS-1 strain of hepatitis A virus. *Infect Immun* 14:1000–1003, 1976.

20. Blumberg BS, Alter HJ, Visnich S: A "new" antigen in leukemia sera. *JAMA* 191:541–546, 1965.

21. Prince AM: An antigen detected in the blood during the incubation period of serum hepatitis. *Proc Natl Acad Sci USA* 60:814–821, 1968.

22. Giles JP, McCollum RW, Berndtson LW, et al: Viral hepatitis: Relation of Australia/SH antigen to the Willowbrook MS-2 strain. *N Engl J Med* 281:119–122, 1969.

23. Bayer ME, Blumberg BS, Werner B: Particles associated with Australia antigen in the sera of patients with leukemia, Down's syndrome and hepatitis. *Nature* 218:1057–1059, 1968.

24. Hirschman RJ, Shulman NR, Barker LF, et al: Virus-like particles in sera of patients with infectious and serum hepatitis. *JAMA* 208:1667–1670, 1969.

25. Dane DS, Cameron CH, Briggs M: Virus-like particles in serum of patients with Australia-antigen-associated hepatitis. *Lancet* 1:695–698, 1970.

26. Gocke DJ: Extrahepatic manifestations of viral hepatitis. *Am J Med Sci* 270:49–52, 1975.

27. Trepo CH, Thivolet J: Hepatitis associated antigen and periarteritis nodosa (PAN). *Vox Sang* 19:410–411, 1970.

28. Kohler PF, Cronin RE, Hammond WS, et al: Chronic membranous

glomerulonephritis caused by hepatitis B antigen-antibody immune complexes. *Ann Intern Med* 81:448–451, 1974.

29. Brzosko WJ, Nazarewicz T, Krawczynski K, et al: Glomerulonephritis associated with hepatitis B surface antigen immune complexes in children. *Lancet* 2:477–481, 1974.

30. Krugman S, Giles JP, Hammond J: Viral hepatitis, type B (MS-2 strain): Prevention with specific hepatitis B immune serum globulin. *JAMA* 218:1665–1670, 1971.

31. Redeker AG, Mosley JW, Gocke DJ, et al: Hepatitis B immune globulin as a prophylactic measure for spouses exposed to acute type B hepatitis. *N Engl J Med* 293:1055–1059, 1975.

32. Surgenor DMN, Chalmers TC, Conrad ME, et al: Clinical trials of hepatitis B immune globulin. *N Engl J Med* 293:1060–1062, 1975.

33. Seeff LB, Zimmerman HJ, Wright EC, et al: Efficacy of hepatitis B immune serum globulin after accidental exposure. *Lancet* 2:939–941, 1975.

34. Krugman S, Giles JP, Hammond J: Hepatitis virus: Effect of heat on the infectivity and antigenicity of the MS-1 and MS-2 strains. *J Infect Dis* 122:432–436, 1970.

35. Krugman S, Giles JP, Hammond J: Viral hepatitis, type B (MS-2 strain): Studies on active immunization. *JAMA* 217:41–45, 1971.

36. Krugman S, Giles JP: Viral hepatitis, type B (MS-2 strain): Further observations on natural history and prevention. *N Engl J Med* 288:755–760, 1973.

37. Barker LF, Chisar FV, McGrath PP, et al: Transmission of type B viral hepatitis to chimpanzees. *J Infect Dis* 127:648–662, 1973.

38. Purcell RH, Gerin JL: Hepatitis B subunit vaccine: A preliminary report of safety and efficacy tests in chimpanzees. *Am J Med Sci* 270:395–399, 1975.

39. Hilleman MR, Buynak EB, Roehm RR, et al: Purified and inactivated human hepatitis B vaccine: Progress report. *Am J Med Sci* 270:401–404, 1975.

40. Buynak EB, Roehm RR, Tytell AA, et al: Development and chimpanzee testing of a vaccine against human hepatitis B. *Proc Soc Exp Biol Med* 151:694–700, 1976.

41. Maupas P, Goudeau A, Coursaget P, et al: Immunization against hepatitis B in man. *Lancet* 1:1367–1370, 1976.

42. Greenberg HB, Pollard RB, Lutwick LI, et al: Effect of human leukocyte interferon on hepatitis B virus infection in patients with chronic active hepatitis. *N Engl J Med* 295:517–522, 1976.

43. Desmyter J, Ray MB, DeGroote J, et al: Administration of human fibroblast interferon in chronic hepatitis B infection. *Lancet* 2:645–647, 1976.

5

Influenza

EDWIN D. KILBOURNE

Influenza is a disease of paradoxes, and the paradox today is the fact that the disease has finally received the attention it deserves at a time when it seems not to exist. Influenza is many things, but epidemiologically it is essentially two things, again paradoxically: a nuisance disease, a minor inconsequential illness, simply one of a number of respiratory virus infections; and an important disease that can kill. These two dimensions of influenza are part of the problem of its assessment and the assessment of measures taken for its control.

CLINICAL SYMPTOMS AND EPIDEMIOLOGIC CLUES

Clinically, influenza is essentially an acute tracheitis. Usually the pathogenesis involves primary implantation of the virus in an alveolus, but most of the cytopathic replication occurs in the middle respiratory tract, producing the characteristic substernal burning and (again paradoxically) the characteristic systemic manifestations of the disease in the absence of demonstrable viremia (1).

In terms of its clinical identification, the simplest rule of thumb is that it is virtually the only respiratory virus that produces febrile respiratory disease in the adult. This is a simplistic and false generalization, but it is useful, I think, in guiding clinical conduct. Death is occasioned by a compound form of virulence in which bacteria are ordinarily involved. However, the virus itself can cause a fatal primary interstitial pneumonia, usually in patients with compromised cardiorespiratory function.

A further epidemiological point about the impact of influenza is that it is the only known modern-day disease of any consequence, infectious or otherwise, that can be detected by a simple inspection of excess mortality in a community; if one has excess total mortality, influenza is almost invariably present and operative, sometimes even in the absence of manifest epidemic disease in younger people.

Influenza has a time scale (Table 5.1). One cannot really talk meaningfully about influenza without considering the specific context of time of its occurrence. Epidemics of influenza described in 1957 were far more severe than the influenza outbreaks seen 4 years after the introduction of that major subtype variant. In recent years, the introduction of major new subtypes of influenza A virus has been occurring at 11-year intervals. In the decade subsequent to each such introduction, after the initial saturation of the human population with the new major variant (which occurs within a 2-year period), the population

TABLE 5.1. *Classification of Human Influenza Viruses*

Influenza Virus Type	Subtype*	Prevalence	Intrasubtype Antigenic Variant
A	H$_{sw}$1N1 (swine)	1918–1929	Not known
	H0N1	1929–1946	Several
	H1N1 ("A prime")	1946–1957	Several
	H2N2 ("Asian")	1957–1968	Several
	H3N2 ("Hong Kong")	1968–?	Hong Kong†
			England
			Port Chalmers
			Scotland
			Victoria
			(others)
B	None, but antigenic variation occurs		
C	None; significant antigenic variation not established		

* Major antigenic variant associated with pandemic diseases at the time of initial introduction.
† Minor variants of Hong Kong (H3N2) subtype.

develops increasing levels of specific antibody against that new variant. Concordant with the development of antibody is the declining magnitude of epidemics that occur with biennial periodicity.

VIRUS CHARACTERISTICS

Mutations and Implications

During a period of less severe outbreaks, the virus is going through progressive mutations that involve the coat proteins through which it interacts with the human host. A unique feature of this infectious agent is its capability for mutation and particularly for antigenic mutation. It does not maintain antigenic constancy as do enteroviruses, measles viruses, or other viruses for which we have vaccines; the influenza virus is constantly shifting. During the recent period of prevalence of the Hong Kong subtype, these minor variants were seen in the form of the so-called England and Port Chalmers viruses and, more recently, the Victoria strain of 1975–1976.

After about a decade of circulation of the major subtype, two remarkable things happen. One is the abrupt disappearance of the virus that has been circulating throughout the world, and the other is the sudden appearance of another major influenza A virus subtype to take its place and repeat the cycle. Admittedly, this analysis of events leading up to 1976 and Fort Dix, New Jersey, is somewhat simplistic. From recorded observations, experts have concluded that two major influenza A subtype variants cannot circulate concurrently in human beings. Immediate replacement of the vanished virus occurs to fill an apparently vacated ecologic niche.

Designations

Influenza viruses are of three major types, A, B, and C. Only influenza A virus is responsible for pandemic disease, has animal reservoirs, or can infect animals in the nonexperimental situation. Relatively less is known about the epidemiology of B and C. Influenza B virus is an important pathogen, particularly in adolescents. However, work on influenza C virus is being done in relatively few laboratories, one of which is that of Drs. Calderon Howe and Richard O'Callaghan at the L.S.U. School of Medicine. It does appear to be a true influenza virus in the sense that it has a segmented genome, but it lacks the viral neuraminidase that characterizes the A and B types.

The designations used when one talks about the "Hong Kong" or "Port Chalmers" strains are confusing. These are intrasubtype variants of major subtypes for which place names are usually given to indicate the geographic location where the variant was first recognized. Major variants of influenza B and C viruses have not been encountered (Table 5.1).

Influenza cannot be understood without information about the anatomy of the virus. It is an enveloped virus, and as such, it has virus-coded glycoproteins inserted into a lipid envelope. The interior of the particle is bounded by a matrix, or M protein. The exact structural details within the particle are not resolved. However, closely associated with eight segmented pieces of RNA, which are the separate genes of the virus, is a nucleoprotein closely wrapped around the viral RNA. RNA polymerases are also closely associated with the nucleoprotein complex. Exactly where they are located is not clear. The important thing to know is that the two surface antigens with which antibodies react are important in immunity. Furthermore, the implications of having a segmented genome and having genes available for random reassortment may be important in the strategy of the virus for survival.

THE VIRUS IN RETROSPECT

A virus similar to the swine virus was first isolated by Shope in 1930 and was probably the virus of 1918 on the basis of indirect but substantial evidence, a kind of seroarcheology involving study of the antibody levels of older persons. Of course the virus of 1918 was never isolated, but the retrospective analysis of its immunologic footprints is, I think, impressive evidence. Most of those who lived during the period of 1918 through 1929, the time when the virus was circulating, have demonstrable antibody to this agent.

The influenza virus was first isolated from man in 1933. Again retrospectively, by looking at excess mortality, one can reconstruct the year 1929 as the probable time of entry of the H0N1 (Table 5.1) subtype into the human population. The magnitude of antigenic variation that could be expected was first appreciated in 1947, and I think all our attitudes toward vaccine have been colored by that experience in the sense that extensive vaccination of military personnel met with complete and dismal failure at that time. This failure was caused by a new subtype (H1N1) in an H0N1-vaccinated population. Obviously, immunization may be expected to be ineffective when the antigens of the vaccine do not match the challenge or infectious agent. Another notable characteristic of 1947 was that the disease was pandemic only in the sense of widespread distribution of virus in a short time, but it was not associated with a high excess mortality, suggesting that variation may occur not only in the antigenic nature of the virus but also in its primary virulence.

The introduction of the Asian virus in 1957 was really the first clear-cut demonstration that something approaching the 1918 situation in severity might, in fact, be caused by a modern, recognized influenza virus. The mortality in 1957–1958 approached 70,000 in the United States, and the impact of that epidemic was extensive. In 1968, with the advent of the Hong Kong decade, another lesson was learned with realization that the Hong Kong virus had only changed its hemagglutinin antigen, the major antigen, but the neuraminidase antigen was preserved. Therefore, one knew that the minimum requirement for being a pandemic virus was a change in the hemagglutinin. Not surprisingly, the severity of the impact of this mutation was less than that observed in 1957 when both antigens changed considerably.

The 1918 pandemic taught us that the influenza virus was among the prime plagues of man. That experience, with 20 million deaths worldwide, certainly rivaled any of the great plagues of the past in magnitude of impact on the population. The severity of the 1929 epidemic and that of 1946–1947, I think, would have to be graded as relatively low; the epidemic in 1957, at the time of change of both antigens, was much higher and that of the 1968 epidemic with one antigen change, somewhat less. In terms of the expected timetable of 11-year mutations seen in the recent past (1946, 1957, 1968), the next major variant would be expected, not in 1976, but in 1979. In terms of the recycling of antigens that may occur at 68-year intervals, the timing is also a little out of phase for return of the swine virus this year.

Despite the evidence that on a large scale of observation, one can get a hint about virulence changes in the virus, a remarkable observation from the records of 1918, an older epidemic, is the constancy of the clinical expression of disease. A typical case of influenza in 1918 was not a fatal case, but rather it was a three-day fever after which the patient promptly recovered. However, the case/fatality ratio in that epidemic approached 1%, whereas it is less than 0.1% today.

VIRUS VARIABILITY, VIRULENCE

Influenza, then, is a relatively unvarying disease caused by a varying virus. If one looks at the clinical profiles from analyses of patients' symptoms in two different pandemics (to which could be added the 1968 experience), the symptoms are remarkably constant (Table 5.2). Fever and cough almost invariably occur in those cases that we recognize. Of course, 20% of those infected with the virus may have no symptoms at all.

This is an important point to note because much formless discussion is now going on about whether the swine influenza virus is more or less virulent than other strains. We have, unfortunately, no reliable *in vitro* markers for virulence for virtually any virus and certainly not for in-

TABLE 5.2. *Clinical Profiles of Influenza in "Mild" and "Severe" Pandemics*

	Influenza in:	
	1947	1957
Fever	99%	97%
Cough	97%	80%
Headache	86%	63%
Nasal discharge	70%	47%
Sore throat	49%	43%
Myalgia	60%	30%
Mean maximum temperature	39°C	39.3°C

fluenza. Nor can we effectively judge virulence by studies in volunteers, for if volunteers are given virus that has gone through any kind of alien-host passage, the virus may have been genetically changed and therefore is not representative of the original virus isolated from man. In addition to having sufficient virulence to cause people to cough and expel it into the environment, the pandemic virus must be transmissible. Transmissibility and virulence can be shown to be separate and segregable properties by recombinational experiments.

ANIMALS AS VIRUS RESERVOIRS

A remarkable feature of the 1968 episode was the fact that it provided for the first time clear prospective evidence of what presumably had happened in 1918. On circumstantial evidence, it had appeared that virus infection in swine was a new disease in the American Midwest in 1918, which followed introduction of the virus into man, and it looked then as though the virus was descending the phylogenetic ladder into swine. Despite careful surveillance, no evidence has since been found that similar transmission has been happening with the later human strains of influenza virus. Subsequent to 1968, however, with the introduction of the Hong Kong virus, considerable spread of that virus subtype to swine occurred.

This factor may be of critical importance, because one hypothesis concerning antigenic variation, particularly major subtype variation, is that animals are reservoirs of virus and that the recombination and recapture of antigens derives from that source. If a human virus could again invade swine, that process would be facilitated. The 1918 swine virus has persisted in swine to this day. Thus, swine are now infected with two viruses, the 1918 and the post-1968 strains, instead of one, and the opportunities for *in vivo* recombination, therefore, clearly exist.

MAN-VIRUS INTERACTIONS

During the interpandemic period after the introduction of a major variant, both the virus and man adapt to their encounter—man by forming specific neutralizing antibody to the virus and the virus by antigenic mutation to escape the human immunologic response. For reasons that are poorly understood, survival of the virus as an obligate human parasite depends on its capacity for antigenic change and the selection of mutants of different antigenicity. In recent years, such variants have included the Port Chalmers and Victoria strains of the Hong Kong virus. After a relatively brief period (10 to 11 years) of prevalence in the human population, the virus disappears to be replaced by a major antigenic variant that in turn produces pandemic infection.

At the end of the interpandemic period, which we are now approaching (and, as far as I am concerned, have reached with reference to Hong Kong subtype prevalence), you will find persons having a composite of heterotypic and homotypic antibodies directed against the variants expressed in the past few years.

Immunity to the influenza virus is not easy to study. Adults are highly susceptible to invasion in pandemic years. That susceptibility, of course, relates to the fact that childhood-acquired antibodies to influenza infections are inadequate against major antigenic variations. However, some degree of heterotypic immunity occurs, even among the major subtype variants. Clearly, the attack rate is higher in children than in adults, and therefore some degree of natural immunity occurs. Where it has been studied, the immunity to recurrence of the disease also seems to be solid, although some reinfection may occur with variants of the same subtype. The chance of expression of the disease sequentially on two different episodes of infection with the same virus type is remote. That observation is important because it does suggest that vaccination, particularly live-virus vaccination, will have utility for the ultimate control of influenza.

Minor antigenic variation in the hemagglutinin and neuraminidase proteins of influenza virus are readily explained as the result of successive point mutations in the viral RNA. In other words, such variants as the Port Chalmers and Victoria strains are probably direct lineal descendants of earlier Hong Kong strains.

On the other hand, the external proteins of viruses of different subtypes (e.g., Hong Kong and swine) differ considerably as determined either by antigenic analysis or peptide mapping. The magnitude of this difference has led to the concept that the emergence of new subtypes is not readily explicable on the basis of serial lineal descendency from the antecedent types and that one must look elsewhere for an explanation of how such different proteins evolve.

The segmented genome of the influenza virus is a remarkable piece of genetic machinery, one possessed by only a few other viruses (such as reoviruses and plant viruses). If, for example, two different strains of influenza A virus infect the same cell (Fig. 5.1), random reassortment of the eight RNA pieces may occur. Such random reassortment, then, can result in asymmetrical biparental contribution to the progeny with any mixture of genome segments. Such mixing confers on the virus something analogous to the genetic adaptability provided us by bisexual reproduction and thus enables it to go beyond the constraints of classical point mutations for the fabrication of new variants. The mixture of genes occurs on an allelic basis, however. The mixture cannot result in two different hemagglutinins in the same particle, except by nonheritable transient phenotypic mixing.

In examining the surface antigens of the virus (Fig. 5.2), which can be clearly assayed in terms of biologic function, or antigenically, one sees that the neuraminidase antigen can be derived from one parent and the hemagglutinin from the other. These so-called antigenic hybrids have been remarkably useful in serologic techniques devoted to looking specifically at antineuraminidase or antihemagglutinin antibodies.

My viewpoint on the major variants is that the barnyard may, in fact, be the graveyard for old influenza strains of man (2). From limited surveys, investigators know that these involve at least 17 strains of animal influenza virus, which would encompass about 11 different hemagglutinins and 8 different neuraminidases in various combinations. Of course, transfer of virus from animals to man is rare; usually, transmission occurs in the reverse direction. Two things must be ex-

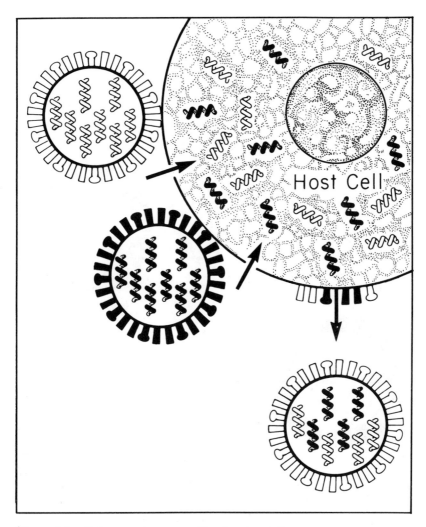

Figure 5.1.   Influenza virus recombination (model).

plained: Whence come the major antigenic variants? And if they come from the animal, how can they come from an animal that so rarely transmits infection to man except, perhaps, by some kind of mechanism for acquiring genes that facilitate this interaction? If one imagines that as a relatively rare event the human virus subtype (known to be only briefly prevalent in the human population) infects a lower animal, as occurred in 1918 (the virus then persisting in swine through the

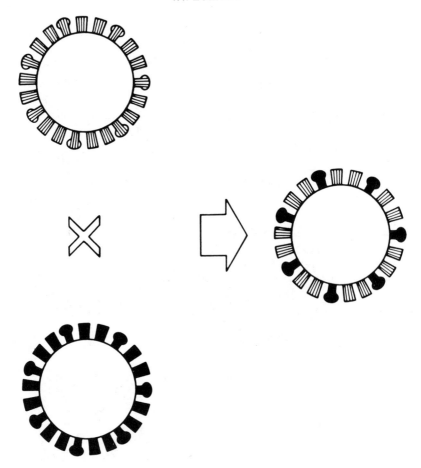

Figure 5.2. Recombination of influenza viruses leading to production of an "antigenic hybrid" containing hemagglutinin antigen derived from one parent and neuraminidase from the other.

period of decline and fall of several human viruses), one can envision, as another rare event, coincident infection of an animal cryptically bearing the subtype that has "disappeared" with the prevalent human virus. The antigenic subtype then might hop the species barrier and be brought back into the human population on the basis of having some genes derived from the human strain.

However, it must be reported that in work from our laboratory, recently published in *Nature,* Palese and Schulman (3) were able to define the genotypes of the Fort Dix virus and other swine influenza viruses on the basis of RNA gel patterns, defining characteristic profiles for different viruses. That work suggests that the Fort Dix swine strain probably was *not* derived by recent recombination in the animal. This finding might have several implications. Possibly, the Fort Dix virus, being more "swinelike," could have more of the potential of the 1918 virus. On the other hand, if the virus has not acquired genes from the prevalent Victoria virus, it may have less ability to spread.

By the mechanism of interspecies recombination and transmission, however, a kind of "instant adaptation" of an antigenically novel virus to man could occur (2). An observation that bears on the subject concerns the reappearance of major antigens seen in the past. One argument about the likelihood of the reexpression of the swine virus as a pandemic virus in 1976 relates to a pattern that, again, has been revealed by serologic archeology. Just as the 1889 pandemic was associated with a virus antigenically similar to that of 1957, so was the virus of 1900 similar to that of 1968. The suggestion here is of a major periodicity in terms of 68-year cycles. If the 68-year cycles are projected from 1918, the time of the swine virus prevalence, one would not expect to see its reappearance, according to a strict timetable, until 1986.

The hemagglutinin and neuraminidase viral antigens are both large proteins. Immunologic studies indicate that the proteins contain multiple antigenic determinants. These proteins must perform biologically and structurally in helping the virus interact with the cell and with antibodies, and the selective pressure for their variation is great. On the other hand, in assessing the reasonableness of any hypothesis that says antigens are limited in their number and therefore must be recycled for the virus to persist, one must appreciate that important structural constraints may limit acceptable variations of these proteins to the extent that they are (a) structurally appropriate for insertion into the envelope of the virus and (b) able to interact as they do with cellular receptors. That the number of useful configurations of hemagglutinin might be limited is not unreasonable. Therefore, I think the concept of recycling has credibility. If it is reasonable, then it attaches perhaps a little more credibility to the tenuous projections we have to make from limited data.

91

FORT DIX AND SWINE INFLUENZA

At Fort Dix, New Jersey, in January 1976, a major variant from the prevalent Hong Kong subtype, namely a swinelike virus, appeared in the recruit population. In association with extensive infection with the Victoria virus (a variant of Hong Kong virus), the swine influenza virus infected about 500 recruits (Table 5.3). Virus was isolated from five patients, and specific antibody response was shown in eight patients. An analysis of these firmly authenticated cases and comparison of their symptoms with those of patients infected with Victoria virus indicated no exceptional virulence of the swine virus, although pneumonic infiltrates occurred in several recruits, and one died of viral pneumonia (4). On these limited data and from what was known from earlier serologic studies, one could conclude the virus had two viral antigens like the putative virus of 1918. Of course, the nature of the other structural proteins of the 1918 virus is unknown.

The other thing learned immediately was that swine influenza virus (which had been transferred only sporadically and rarely to man in the past, with no secondary cases discernible) was now clearly transmissible from human to human and therefore was not a "swine" virus any more but a human virus. Fort Dix had no swine source for this virus; nevertheless, it could be transmitted, could cause disease, and, indeed, resulted in the death of one patient. Although much was made of that death, one must appreciate that all influenza A virus strains can kill people.

Also of epidemiologic importance (and bearing on the fact that we have no swine influenza virus palpably with us at present) is the fact

TABLE 5.3. *Swine Influenza in Man—Data from Fort Dix*

---

1. Major variant from Hong Kong subtype
2. Swine influenza virus—transmission, disease, and death in human population
    Virus isolation—5 cases
    Antibody response—6 cases
    High antibody level to swine influenza virus—500 cases
3. Virulence unexceptional (limited data)
4. Two viral antigens like those in 1918 virus
5. Swine influenza virus disappeared in competition with Victoria strain

---

that, in competition with the Victoria strain, which was operating in a partially immune population, the swine influenza virus disappeared. This fact may indicate limited transmissibility of the swine virus, even when circulating in a population that had virtually no antibody to it.

Another part of the decision whether to make the recommendation for the control of an incipient pandemic for the first time in history involved recognition that we have had in hand, since 1974, an effective immunizing agent in the form of inactivated influenza vaccine (Table 5.4). That vaccine induces a good serologic response and is so effective an antigen that it is used for immunologic studies. The vaccine induces immunity in controlled studies at the level of 70% to 90% protection. Data on the duration of immunity are limited, but it varies from 6 months to 3 years, probably with a median duration of about a year. The reactogenicity of the modern sucrose-density-gradient purified preparation is remarkably low in adults; it is one of the safest immunizing agents that exists. A search of the literature by the Center for Disease Control turned up only three vaccine-associated deaths in a 20-year period (personal communication with M.A.W. Hattwick, C.A. Hoke, Jr., and R.J. O'Brien). The relatedness of two of these to vaccination, I think, is questionable.

Subsequent to presentation of this paper, a temporal relationship between the administration of swine influenza vaccine and the increased risk of occurrence of Guillain-Barré syndrome has been observed by prospective surveillance of more than 42 million persons immunized in the United States within a three-month period. It is unclear at present whether this increased risk of acquiring a rare disease is specifically related to influenza immunization or perhaps to any immunization procedure effected on a mass basis. The risk of Guillain-Barré syndrome ($\cong$ 1 : 140,000 injections) is trivial when the risk of severe and fatal pneumonia in the elderly is considered. In recognition of this fact, the United States Public Health Service has recommended the reinstitution of vaccination of those at high risk for complications of influenza (5).

TABLE 5.4. *Characteristics of Inactivated Influenza Vaccines*

Induce good serologic antibody response
Induce immunity to influenza virus of same variant type (70%–90%)
Duration of immunity—6 mo to 3 yr (mean = 1 yr)
Low reactogenicity (in adults)

The problems encountered through the years in the application of this effective agent have not been wholly scientific. Variable physician and patient acceptance are historical problems because of previous problems with the potency of this vaccine as monitored by the Bureau of Biologics. Reaction rate occasionally has been high, chiefly as a result of bacterial endotoxin. The immunity is not sustained, but that may be a characteristic of other inactivated vaccines as well.

Confusion always occurs about immunization objectives, and, in an interpandemic period, no one has ever seriously considered use of mass vaccinations for influenza with this product because of the brevity of immunity and the problems related to administration (Table 5.5). The immunization target, unlike that of such well-defined diseases as measles or paralytic poliomyelitis, will be confused with targets for other respiratory infections, and hence, confusion can occur if efficacy is judged in the absence of specific serologic investigations of the immunized population. Vaccine supplies have been perennially inadequate: 20 million doses a year for 45 million persons at high risk. Maldistribution results in vaccination of persons who are not exceptional risks. When one adds to this the problem of viral variation and therefore difficulties in predicting when influenza will appear, the poor utilization of the vaccine does not come as a surprise.

The immunization objective in a pandemic (something never anticipated before) is to use this vaccine, which admittedly produces briefly sustained immunity, for blunting that *initial impact* of the epidemic with which maximum mortality and untenable morbidity are associated, the latter usually involving 30% of the population. Thus the objective changes from reduction in mortality alone to prevention of illness. The wholesale morbidity of influenza pandemics should no longer

TABLE 5.5.   *Immunization Objectives*

Interpandemic period	Primary Prevention of mortality in disease with case fatality ratio of $<0.1\%$ Secondary Prevention of morbidity in some (recommendations mixed and variable)
Pandemic period	Prevention of mortality (case/fatality ratio usually $<0.1\%$, but there are many cases) Prevention of wholesale morbidity

be tolerated in the 20th century (Table 5.5). Also, in the pandemic, mortality expands dramatically, not because the case/fatality ratio is necessarily any greater and not because of any intrinsically greater virulence of the virus, but simply because a great many persons are infected, and at the same percentage of fatality, many die.

SCENARIO FOR 1977–?

About 200 million doses of vaccine are now at hand, and we have, at this writing, no epidemics of the target disease. The answers are not in and they will not be in until the winter of 1976–1977 is past. Influenza is a wintertime disease, but it is not necessarily a wintertime disease at the time of a major pandemic introduction. If one conceives of the swine virus as being a virus that had trouble getting going even in a recruit population and if one also assumes that infection of 500 recruits did not reach a dead end, then one may assume that some human foci must have existed for the importation of virus into Fort Dix to begin with. Prospective serologic surveillance has shown that every year in the United States, a silent chain of influenza A virus transmission occurs, which is detectable in July as well as in January. Therefore, it is not unreasonable to believe the swine influenza virus now remains somewhere within the human population. When the proper seasonal circumstances, which include low relative humidity and crowded environments come into play, we may well see its reemergence.

The script can be written in several ways, and it has always been recognized that the pandemic might not occur at all. Nevertheless, every time previously that a subtype (major) variation has been recognized, it has been followed by pandemic disease. Several things can happen. The pandemic may, indeed, not appear in 1976. However, if it does appear in January or February 1977 (the influenza months), one can then conclude that an animal virus can indeed cause a human pandemic and may be recognized before the event. If the Palese and Schulman data (3) on the viral genotype can be taken at face value, then recombination may not be necessary for interspecies transfer of virus in pandemic form.

We have already learned from this experience that concurrent circulation of major subtypes *can* occur in man. If swine virus does *not* reappear in man, one can conclude that a major variant can emerge in a susceptible population and *not* produce a pandemic in that population. This emergence without pandemic may, of course, have occurred before but escaped detection. If the virus, on the other hand, reappears later—after 1977 and perhaps more in accord with the expected timetable of events—then identification of a pandemic strain *before* its general distribution and recognition will have been shown. Perhaps, then, the 1976 experience with swine virus would corroborate the belief that good surveillance might have detected the Asian variant on the Chinese mainland in 1956 before it began to circulate widely in 1957.

Influenza continues to baffle and perplex its students and to frustrate those who strive for its control. But we must not abandon the effort because of the magnitude of the task.

DISCUSSION

Question: No mention has been made of gastrointestinal (GI) symptoms. In the 1968 influenza epidemic, large numbers of patients in northern India had nausea, vomiting, and diarrhea followed by respiratory symptoms. What should one expect as GI symptoms in viral infections?

Dr. Kilbourne: I think in instances where the disease has been carefully studied by means of virus isolation and serologic response and also adequately characterized clinically, few GI symptoms are noted. Children may show some GI symptoms but nonspecifically in relation to febrile disease. Influenza virus, of course, does not replicate in the GI tract, only in the pharynx and the trachea, so that the expression of symptoms related to the GI tract might not be expected. Unfortunately, the clinical nomenclature includes the term, "intestinal flu," which is a misnomer, and probably these effects are more related to the rotaviruses that Dr. Melnick described.

Dr. Krugman: It's possible that the situation may be analogous to that seen with measles. For example, in the United States, the clinical syndrome of measles is not characterized by gastroenteritis and diarrhea. On the other hand, in developing areas of the world, measles is characterized by a high fatality rate frequently associated with severe diarrhea and dehydration. Whether the gastroenteritis is due to measles virus per se or whether it is the endemic rotavirus or a bacterial pathogen is not known, but many infants in Nigeria, for example, frequently have diarrhea associated with measles. Conceivably, this relationship could occur with influenza in some parts of the world.

Question: Since the consumption of the swine virus vaccine does not seem to be great, can it be held over until next year in case we need it? Also, has some difficulty occurred in having an ideal amount of the hemagglutinin and neuraminidase antigens in the product?

Dr. Kilbourne: Distribution of vaccine is always a problem. I think this is a reason that the authorities set up a national program: to ensure that the appropriate priorities are set in terms of immunizing, first, those with a so-called high risk of the severe consequences of infection and, later, those who are at less risk of serious disease. The problem is a delicate one involving public relations, national and local politics, and the like. Looking back over the things that have happened with this program, one can separate out the scientific problems such as those that relate to neuraminidase antigen from those that are nonscientific in relation to events, some of which have been foreseen and some not. It does seem clear on epidemiologic grounds that no vaccine-associated deaths have occurred as yet, but program-associated deaths have occurred. Of course, when one immunizes persons selectively who are elderly or otherwise at high risk, it is not unexpected that some will die, as some did as they were reading the consent form. Such deaths can be construed as being vaccine-associated in the broadest sense, but they certainly cannot be linked to the pharmacology of the vaccine.

Other problems of distribution are going to be far more important, and these relate to the immunization of children, an important target population. With improved studies of virus distribution, we recognize increasingly the importance of influenza in the very young. The problem is that, as has been foreseen, the influenza virus is toxic in children, except in restricted dosage. However, careful trials conducted during the summer have shown that it is possible to arrive at a dose regimen that is acceptable and sufficiently immunogenic in the childhood population. The misfortune is that this involves administering two small doses of vaccine. Because of the difficulties in administering even one dose of vaccine into the public, I think this program has the seeds of defeat in it. But the Public Health Service is obligated to recommend something covering the state of affairs and this they will do soon. Apparently the production goal of almost 200 million doses will have been realized by the end of January, but the critical problem now of public acceptance is a real one. If one case of swine flu appeared, I think we would see an immediate reversal of the public's attitude that would override any concerns for vaccine toxicity or the like. Part of the public misinformation about this vaccine, which to me is incredible, is the fact that many physicians regard it as an ineffective *live* virus vaccine, despite the fact it says "inactivated" on the label. The implications are different for the two types; one would not hesitate to immunize pregnant women, for example, with the inactivated vaccine.

REFERENCES

1. Kilbourne ED (ed): *The Influenza Viruses and Influenza.* New York, Academic Press, 1975.
2. Kilbourne ED: The molecular epidemiology of influenza. *J Infect Dis* 127:478, 1973.
3. Palese P, Schulman JL: RNA pattern of "swine" influenza virus isolated from man is similar to those of swine influenza viruses. *Nature* 263: 528–530, 1976.
4. Gaydos JC, Hodder RA, Top FH Jr, et al: Influenza A swine at Fort Dix, NJ (January–February 1976): I. Case finding and study of cases. *J Infect Dis,* submitted for publication.
5. *Morbidity & Mortality Weekly Report,* 26:52 (Feb 11) 1977.

6

Rubella

GILBERT M. SCHIFF

Rubella has come into renewed prominence primarily because of the inherent teratogenicity of the causative virus. The isolation of rubella virus came about just in time to be of special importance in the analysis of the epidemic of rubella that occurred in 1964–1965. Major findings included a precise delineation of the immune response to rubella virus, the feasibility of active immunization, and the recognition of the "rubella syndrome," with all its devastating primary and secondary consequences. Although many basic and practical questions about rubella remain unanswered, or at best only partially clarified, great progress has been made in controlling the disease and hence in reducing the fetal wastage resulting from naturally acquired intrauterine infection. A danger that now threatens is that the means currently available for the control of rubella will be incompletely or inadequately applied and that, as a result, a resurgence of rubella may occur in the next decade. The practicing pediatrician is the bulwark against this danger.

Rubella usually occurs as a mild disease of childhood and is characterized by a three-day maculopapular rash and cervical lymphadenopathy. This disease assumed serious implications in 1941, when Gregg first correlated the occurrence of fetal congenital malformations with maternal rubella contracted during early pregnancy (1). Gregg's observations were rapidly confirmed and eventually led to the realization that during the 1964 nationwide epidemic of rubella in the United States, an estimated 30,000 to 50,000 infants had been affected. Because of the teratogenic effect of the rubella virus, prevention of the disease became a major goal. The causative agent, long thought to be a virus, was finally isolated in 1962 (2–4). This breakthrough was followed by the development of techniques for accurate laboratory diagnosis (5) and eventually the production of several rubella vaccines (6–8). The tools are now available to define the disease and to attempt its prevention. What then is the current status of rubella, its diagnosis and treatment?

ACQUIRED RUBELLA

The present situation with regard to acquired rubella can be summarized in the following statements:

- If available laboratory tests are not used, clinical diagnosis is unreliable in an individual case.
- Individual immunity or susceptibility to rubella can be reliably determined.

101

- The ratio of subclinical infections to clinical infections is at least one (9).
- During active rubella infection, the virus is present in and is shed from the nasopharynx for 7 to 10 days before and up to a week after clinical signs become evident (4).
- Viremia may occur as early as 6 days after exposure but disappears when signs of the illness are manifest (4).

In making a clinical diagnosis of rubella, the clinician may be misled in two ways. First, rubella can occur either without the characteristic rash or with an atypical rash. Second, several unrelated viruses, none of them known to be teratogenic, can cause an illness closely resembling rubella clinically. In a study at the University of Cincinnati, definitive laboratory examinations were done on 24 adults who had been diagnosed clinically as having rubella. In only 6 of those patients was rubella confirmed by virus isolation or serologic tests or both. Fifteen patients were shown to have enteroviral disease and 3, disease caused by unidentified viruses. In 2 patients in whom no diagnosis was established, rubella was ruled out by failure to isolate the virus and by negative serologic tests. In 4 patients who were pregnant at the time of acute illness, 3 had an enteroviral infection; thus, unwarranted termination of pregnancy was avoided by careful laboratory diagnosis. In a study undertaken during the 1964 epidemic of rubella, three nursing students developed an illness that was similar to rubella but was diagnosed by laboratory studies as having been caused by ECHO-9 virus (9).

Rubella can be diagnosed in the laboratory by isolation of the virus or by serologic tests. The virus can be recovered from the nasopharynx up to a week after the onset of symptoms, the greatest amount of virus being shed while symptoms are present. The serologic diagnosis depends on detection of a significant difference in antibody levels between acute and convalescent serum specimens (see Chapter 14). Two types of antibody assays are available, namely, hemagglutination inhibition (HI) (5) and complement fixation (CF) (10) tests. The HI test is used when an acute serum specimen can be obtained relatively early in the clinical illness, because the HI antibody appears soon after onset and rises to a peak in three to four weeks. The CF antibody rises a week or two after onset of symptoms, so that if the patient is first seen even a week or 10 days after signs and symptoms have disappeared, it may still be possible to demonstrate a rise in CF antibody. The HI antibody probably persists for life, whereas the CF antibody will disappear within 6 months to a year after the acute infection. For this reason, the HI antibody test is the more reliable for determining the immune status of a person in the absence of known clinical disease.

Immunity to rubella is closely correlated with the presence of HI antibody, although opinions differ on the critical levels of HI antibody necessary to confer immunity. In our laboratory, the presence of any antibody detectable by HI is considered indicative of immunity. This hypothesis has been tested and proved valid as a result of an experimental challenge with rubella viruses in volunteers (11) with or without detectable serum antibody.

A question has arisen as to whether there are significant differences among strains of rubella virus. For instance, epidemiologic evidence by Kono (12) has suggested that strains of rubella virus recovered in Japan have less teratogenic potential than strains recovered elsewhere. Correlative studies on women who developed rubella during pregnancy showed that the incidence of congenital abnormalities in infants coming to term was less than expected.

Evidence for a possible second strain of rubella virus was obtained during an outbreak of typical erythema infectiosum (fifth disease) in which multiple isolations of rubella virus were obtained from patients with typical disease (13). One such isolate was grown out, safety tested, and administered to men susceptible to rubella. The resulting clinical disease was not typical of fifth disease. It differed from that produced in previous challenge studies with standard strains of rubella virus (14) in that cough was present and the rash that developed was pruritic and of significantly longer duration than that resulting from infection with standard strains of rubella virus; all infected subjects developed a significant level of HI antibody to standard rubella antigen. Two volunteers who failed to develop clinical disease did not develop detectable antibody to standard rubella virus, but they had low levels of neutralizing antibody to the challenge strain of virus. The latter had been shown by Ouchterlony tests to be antigenically identical to the standard strain of rubella virus.

The effectiveness of immune globulin in the prevention of rubella has been the subject of controversy largely because interpretation of results has varied as a result of factors such as time of administration, immune status of recipients, and the availability of specific serodiagnostic tests. The compelling indication for immune globulin is for the pregnant woman who has been exposed to recognized rubella and is found to be susceptible by serodiagnostic tests. Apparently, immune globulin, when given before illness occurs, can modify the disease without, however, any certainty that viremia, and hence fetal infection, will be prevented. From a number of studies, the time from exposure to the appearance of viremia has been estimated at six days. To prevent viremia, therefore, immune globulin must be given within six days of exposure.

A study was conducted at the University of Cincinnati to determine whether administration of immune globulin of high titer at an optimal time (24 hours after exposure) would prevent viremia (15). In susceptible men, intramuscular administration of 20 ml high-titered (2048) immune globulin completely prevented not only rubella infection but also viremia under these conditions. Lower-titered immune globulin (256) was ineffective in preventing viremia. Proof that the former group had been passively protected against the first viral challenge came when they were challenged with rubella virus a second time six months later and developed the characteristic clinical illness, confirmed serodiagnostically. On the other hand, L. Cooper, in a personal communication, reported failure to protect children by giving immune globulin in similar dosage before challenge with the same amount of virus. This discrepancy may be due to the relatively greater impact of the same viral challenge in the younger age group.

At the University of Cincinnati, our policy has been to administer 20 ml of immune globulin to the serologically susceptible pregnant woman who is exposed to rubella but under no circumstances wants to terminate her pregnancy. The patient must agree to have serial blood tests to detect subclinical rubella infection.

CONGENITAL RUBELLA

The situation with regard to congenital rubella may be summarized in the following statements (16):

• Additional abnormalities have been recognized recently as being associated with congenital rubella infection such as hepatitis, pneumonitis, myocarditis, lesions of the long bones, thrombocytopenic purpura, and hemolytic anemia.

- Abnormalities of the fetus can result from subclinical maternal infection.
- Infants born with congenital rubella shed virus for long periods.
- Infants with congenital rubella have been shown to be vectors of rubella virus.
- Immunologic tolerance does not underlie congenital rubella infection.

The foregoing facts were determined from results of laboratory studies performed during the 1964 nationwide epidemic of rubella. In Cincinnati, 115 cases of congenital rubella were virologically and serologically identified among the more than 4000 births that occurred from the fall of 1964 to October 1965. Of the infants so identified, 75% appeared to be perfectly normal in the neonatal period (17). Follow-up examination of some of this group of 115 patients revealed additional abnormalities that became evident only during later development. However, many of the children still appear to be normal at the beginning of the second decade of life and may be said, therefore, to have suffered "subclinical" rubella syndrome. However, the syndrome of congenital rubella has recently been amplified by the recognition of several cases of subacute sclerosing panencephalitis similar to that associated with rubeola (18,19). Those cases that occurred in children with known congenital rubella had their onset in the second decade and were associated with persistently high levels of serum and cerebrospinal fluid antibody to rubella virus. An agent resembling rubella virus was isolated from the brain tissue of one patient. These cases may be a premonition of serious sequelae among the many thousands who were born in 1964–1965 with congenital rubella and who are now in their second decade of life. That chemoprophylaxis may lessen the impact of possible SSPE in these patients has been suggested. Amantadine was shown to inhibit growth of rubella virus *in vitro* and, in an infant with congenital rubella, to reduce temporarily the amount of rubella virus being shed (20). In children known to have had congenital rubella, persistently high levels of serum and cerebrospinal fluid antibody to the virus might be considered as an indication for amantadine prophylaxis against the development of SSPE. Amantadine therapy might even be considered in children in whom SSPE is already clinically apparent (see Chapter 18).

In the overall immunologic response to congenital rubella infection, the pattern for humoral immunity has been well defined. The fetus responds by producing IgM HI antibody, which then is present at birth and is soon replaced by IgG antibody. Significant levels of IgG HI antibody persist for several years and then tend to decline. Attempts to boost the level of IgG antibody by vaccination have failed (L. Cooper, personal communication).

The pattern for cell-mediated immunity in rubella is less clear, particularly in view of the conflicting data that have been reported regarding deficiencies of cell-mediated immunity in congenital rubella (16).

The development of an animal model system in which to study congenital rubella is an important goal and would provide an opportunity to elucidate teratogenic mechanisms and effects of virulent virus and to evaluate the teratogenic potential of attenuated rubella vaccines. To this end, we have demonstrated in our laboratory that transplacental rubella virus infection can be achieved in ferrets and that the virus can be recovered from the tissues of full-term offspring of mothers infected early in the gestational period. The ferret seems a particularly attractive model for the study of human congenital rubella because its placentation is similar to that in humans, the gestational age is easy to measure, and ferrets are susceptible to infection with rubella virus.

RUBELLA VACCINES

Several live, attenuated rubella virus vaccines have been licensed in the United States. Virus for the HPV77DE5 vaccine was attenuated by multiple passages in monkey kidney cells followed by passages in duck embryo cells; HPV77DK12 vaccine was attenuated by multiple passages in monkey kidney cells but followed by passage in dog kidney cells; and the Cendehill strain was passaged in rabbit kidney cells. The vaccine grown in canine kidney was removed from the market because it was associated with a high rate of reactions. A fourth vaccine (RA27/3), licensed in England and France, was developed by passage in human embryonic lung cells (WI38) and will probably soon be licensed in the United States. Immunization in the United States has been directed toward children from one year of age to puberty in the hope that sufficient "herd immunity" would thereby be developed to protect the real target of prophylaxis, the susceptible pregnant woman. Because of theoretical teratogenicity of rubella vaccines, they should not be administered to pregnant women.

Since the vaccines became available in 1969, the nationwide reduction in the number of cases of rubella has been dramatic. In Cincinnati, before the availability of the vaccines, 150 to 250 confirmed cases of rubella were seen each year. In 1970, 85,000 elementary school children in the city were vaccinated, and since then, physicians have been administering the vaccines routinely to infants one year of age. Since 1970, from 3 to 20 cases of confirmed rubella have been seen each year. In 1968, a serologic survey of children just entering school showed that 75% were susceptible to rubella. In 1974, a second survey showed that only 20% of children at the same age were susceptible (21,22). The virtual disappearance of congenital rubella syndrome from the Cincinnati scene constitutes another gratifying result of the vaccination program. Despite these advances, outbreaks of rubella have still occurred in the teenage populations of communities in which the younger children had been vaccinated (23). Moreover, recent surveys have indicated that 30% to 40% of teenage and young adult females remain susceptible to rubella (24). Careful selection and immunization of known susceptible postpubertal females should therefore be strongly considered. A program along these lines has been in operation in Cincinnati and has included immunization of susceptible teenage girls who have already experienced unwanted pregnancies (25). During the past five years, all prenatal patients seen at the Cincinnati Health Department Clinics have been tested for rubella antibody. The susceptibility rate in that group has ranged from 12% to 17%. Those found susceptible are followed carefully through pregnancy for serologic or clinical evidence of rubella infection and are then vaccinated in the immediate postpartum period.

An important question unanswered is the duration of immunity that follows vaccination. The routine administration of rubella vaccines to young children as now advocated makes mandatory the serologic surveillance of vaccine-induced antibody so that those whose antibody levels have waned may be revaccinated. Otherwise, females may be deprived of the long-lasting immunity that would follow natural infection during childhood without any certainty that vaccine-induced immunity would persist into the childbearing years. The early data on that question are somewhat conflicting. In a longitudinal study of 1500

107

children vaccinated in 1968 with either canine or rabbit kidney vaccine, Rauh and I found that only 4% had lost detectable antibody levels in the intervening seven and one half years. Horstmann has reported a much higher percentage who lost antibody within five years after receiving duck embryo vaccine (26). Hilleman and associates (27) found a loss in antibody in 2% of vaccinees five years after immunization with duck embryo vaccine. The critical point here is that surveillance of vaccinees must be maintained, and proper steps taken if immunity is found to disappear with time.

RUBELLA CONTROL

I believe that an effective program for rubella control can be instituted community-wide and that vaccination is only a part of such a program, which would include:

- Routine use of rubella vaccines in prepubertal children.
- Continuous laboratory surveillance of vaccinees to determine the duration of vaccine-induced antibody with booster inoculations of vaccine if antibody levels wane.
- Routine testing of classes of children entering school to identify susceptible groups who then should be immunized.
- Routine testing of females in high school for rubella antibody, as well as premarital and prenatal testing.
- Selection, education, and immunization of postpubertal and postpartum females shown to be susceptible.
- Laboratory confirmation of all suspected cases of rubella.

Such a program would provide continuous evaluation of vaccine efficacy as well as identification of those needing vaccination. In addition, the program would accurately determine the incidence of rubella in the community.

DISCUSSION

Question: Dr. Krugman, how does Dr. Schiff's experience compare with yours?

Dr. Krugman: Many of us who were involved with the traumatic rubella experience in 1964, including Dr. Schiff and his colleagues and
many other investigators throughout the nation, realized at that time
that if and when a vaccine became available, it would meet a definite
need. We were as convinced that we needed a vaccine for rubella as we
were about a need for a poliomyelitis vaccine. If one had the opportunity to follow up infants the way Dr. Schiff did and the way Dr. Cooper
of our staff did in New York City, where he still is observing well over
500 severely damaged infants born to mothers who had rubella, one
would know the problem is real.

Since the licensure of rubella vaccine in 1969, I think we've made
extraordinary progress in the United States. In the past, epidemics of
rubella occurred at six- to nine-year intervals. However, in the wake
of the use of about 70 million doses of rubella vaccine, there has been
a progressive decline in the number of reported cases of rubella and
congenital rubella. The epidemic that should have occurred in the
United States between 1970 and 1974 was prevented. In my opinion,
and I'm sure Dr. Schiff agrees with me, the break in the cycle is related
to the extensive use of rubella vaccine in children who disseminate the
virus.

One big problem in the United States and in Great Britain is that of
trying to immunize women of childbearing age. Although Dr. Schiff
and his group are immunizing such women in Cincinnati, in most communities in the United States it is difficult to convince obstetricians
and internists to immunize adults against rubella. Physicians do test for
rubella antibody to determine susceptibility, but unfortunately a small
percentage of women actually receive the vaccine. In Great Britain,
only prepubertal girls and women of childbearing age were supposed to
get the vaccine. At a meeting I attended there last April, it was reported that under the existing conditions, only 72% of prepubertal
girls and few women of childbearing age were being reached. The reports at that meeting indicated that they were still seeing as much congenital rubella in 1974 and 1975 as they saw before. Contrariwise, the
surveillance data in the United States are similar to the findings that
Dr. Schiff has seen in Cincinnati.

The design of the British program was such that they didn't expect to
see any change in the incidence of congenital rubella for at least 10
years—by 1980 or later. However, they are not reaching as many girls
as they wanted to and, I think unfortunately, that is what's happening

in the United States, too. In retrospect, I believe the decision here was correct, and I believe that the findings that you heard reported by Dr. Schiff of Cincinnati have been observed in many parts of the United States. Recent studies in Hawaii, for example, have confirmed the findings Dr. Schiff presented—more than 98% of about 4000 or 5000 children who had been immunized four or five years ago still have detectable rubella antibody.

Question: How expensive is the testing of high school girls for immunity or antibody to rubella? Can the testing be done on a wide scale? Also, is the rubella virus vaccine a live vaccine, and if it is live, why doesn't it produce lifelong immunity?

Dr. Schiff: The cost per test, I'm sure, varies from laboratory to laboratory. In our laboratory the cost is $2.75 for testing each person, but we do it on a large scale so that's probably cheaper than costs at other places. The vaccine has live attenuated viruses in it. We really don't know yet whether the vaccine does produce lifelong immunity. We've only been able to follow up the children inoculated on a large scale for 7 years now. Drs. Hartman and Meyer have followed up a handful of children for 9 and 10 years now, but whether the vaccine produces lifelong immunity in the majority of the recipients is still unknown. Dr. Horstmann has referred to the difference in initial reaction. Those who do react, react well, but whether or not they will have lifelong immunity remains to be proved. Finally, determination of immunity can be done on a wide-scale basis.

Question: What risk has the rubella vaccine for the pregnant woman who is inadvertently vaccinated early in pregnancy; that is, during the first month or two?

Dr. Schiff: I think the problem of vaccinating a pregnant woman still remains theoretical. A number of women shown to be susceptible to rubella have been vaccinated. They either had their pregnancies terminated or have delivered at term. The information on these products of conception showed that indeed the virus in the vaccine can get across the placenta to the developing fetus, but there is no information that it can cause damage. The same findings occurred for those women who had gone to term and whose offspring were studied. In Cincinnati, a woman gave birth to a baby with Polan syndrome—agenesis of the pectoral muscles, webbing, and cleft palate. The infant died at about four days of age and, in getting the history, the resident found that the woman had received rubella vaccine early in her pregnancy. Her physician had not asked her if she was pregnant at that time; he just gave her the vaccine. We studied the baby, trying to isolate the virus, which can be done easily in natural congenital rubella, but we were unable to do so. We were unable to recover virus on autopsy specimens, which, again, is easy to do. The infant did have antibody to rubella, but it was all IgG antibody with no evidence of IgM. We believe the vaccine was not implicated at all. A year later, the pathologist at the hospital called me down to look at a case of encephalitis in a newborn infant which he said looked just like the rubella encephalitis. We took out the specimens again and tried to isolate the virus and still were unable to do it. The Polan syndrome, as far as I'm aware, has never been associated with natural congenital rubella.

Question: Does fifth disease occur in epidemics?

Dr. Schiff: Yes.

Question: Is the pregnant female at risk from the child?

Dr. Schiff: Not from a recently vaccinated child.

Question: Is high-titer immune globulin available, and what is the titer of the immune globulin you administer to children and pregnant women?

Dr. Schiff: No. We give 20 ml of commercially available immune globulin to pregnant women. The titer is 512-1024.

REFERENCES

1. Gregg NM: Congenital cataract following German measles in the mother. *Trans Ophthalmol Soc Aust* 3:45–46, 1941.
2. Parkman PD, et al: Characteristics of an agent recovered from recruits with rubella, abstracted. *Fed Proc* 21:466, 1962.
3. Weller TH, Neva FA: Propagation in tissue culture of cytopathic agents from patients with rubella-like illness. *Proc Soc Exp Biol Med* 111:215–225, 1962.
4. Sever JL, Schiff GM, Traub RG: Rubella virus. *JAMA* 182:663–671, 1962.
5. Stewart GL, Parkman PD, Hopps HE, et al: Rubella virus hemagglutination-inhibition test. *N Engl J Med* 276:554–557, 1967.

6. Buynak, EB, Larson VM, McAleer WJ, et al: Preparation and testing of duck embryo cell culture rubella vaccine. *Am J Dis Child* 118:347–354, 1969.

7. Musser SJ, Hilsabeck LJ: Production of rubella virus vaccine: Live, attenuated in canine renal cell cultures. *Am J Dis Child* 118:355–361, 1969.

8. Huggelen C, Sigel M, Zypraich N, et al: Safety testing of rubella virus vaccine (Cendehill strain): Preparation in primary rabbit kidney cells. *Am J Dis Child* 118:362–366, 1969.

9. Schiff GM, Smith HD, Dignan P, et al: Rubella: Studies on the natural disease: The significance of antibody status and communicability among young women. *Am J Dis Child* 110:366–369, 1965.

10. Sever JL, Huebner RJ, Castellano GA, et al: Rubella complement fixation test. *Science* 148:385–387, 1965.

11. Schiff GM, Donath R, Rotte TE: Experimental rubella studies. *Am J Dis Child* 118:269–274, 1969.

12. Kono R: Antigenic structures of American and Japanese rubella virus strains and experimental vertical transmission of rubella virus in rabbit. International Symposium on Rubella Vaccines, London, 1968. *Symp Series Immunobiol Standard* 11:195–204, 1969.

13. Balfour HH Jr, May DB, Rotte TE, et al: A study of erythema infectiosum: Recovery of rubella virus and echovirus-12. *Pediatrics* 50:285–290, 1972.

14. Schiff GM, Linnemann CC Jr, Balfour HH Jr, et al: Challenge study with rubella virus isolated from a patient with erythema infectiosum, abstracted. *Clin Res* 19:675, 1971.

15. Schiff GM: Titered lots of immunoglobulin: Efficacy in the prevention of rubella. *Am J Dis Child* 118:322–327, 1969.

16. Schiff GM: Rubella, in Eickhoff TC (ed): *Practice of Medicine.* Hagerstown, Md, Harper & Row, 1971, vol 4, chap 21.

17. Schiff GM, Sutherland JM, Light IJ, et al: Studies on congenital rubella: Preliminary results on the frequency and significance of presence of rubella virus in the newborn and the effect of gamma globulin in preventing congenital rubella. *Am J Dis Child* 110:441–443, 1965.

18. Townsend JJ, Barringer JR, Wolinsky JS, et al: Progressive rubella panencephalitis. *N Engl J Med* 292:990–993, 1975.

19. Weil ML, Itabashi HH, Cremer NE, et al: Chronic progressive panencephalitis. *N Engl J Med* 292:994–998, 1975.

20. Schubert W, Schiff GM, West C: Amantadine therapy in a case of dysgammaglobulinemia type 2 and enteropathy associated with congenital rubella, abstracted. *Clin Res* 14:439, 1966.

21. Schiff GM, Rauh JL, Rotte T: Rubella vaccine evaluation in a public school system. *Am J Dis Child* 118:203–208, 1969.

22. Schiff GM, Linnemann CC Jr, Shea L, et al: Serological survey for rubella and measles antibody among first graders. *JAMA* 227:49–52, 1974.

23. Klock LE, Rachelfsky GS: Failure of rubella herd immunity during an epidemic, *N Engl J Med* 288:69–72, 1973.

24. Schiff GM, Linnemann CC Jr, Shea L, et al: Rubella surveillance and immunization among college women. *Obstet Gynecol* 43:143–147, 1974.

25. Rauh JL, Schiff GM, Johnson LB: Rubella surveillance and immuni-

zation among adolescent girls in Cincinnati. *Am J Dis Child* 124:71–75, 1972.

26.   Horstmann DM: Controlling rubella: Problems and perspectives. *Ann Intern Med* 83:412–417, 1975.

27.   Weibel RE, Buynak EB, McLean AA, et al: Long-term follow up for immunity after monovalent or combined live measles, mumps and rubella virus vaccines. *Pediatrics* 56:380–387, 1975.

7

Encephalitis

JAMES P. LUBY

With the possible exception of the viral hepatitides, no viral diagnosis is more terrifying to the lay public than that of the "sleeping sickness," encephalitis. The public's response is well-founded in light of the epidemic figures compiled for the last calendar year (1975) in which all data are complete. In that year a sizable increase in infectious encephalitis occurred, that is, more than 4000 cases in the United States, nearly tripling the case rate per 100,000 population when compared with the 1974 figures (1.80 versus 0.50/100,000). More than 2000 cases of laboratory-documented arbovirus infections alone were reported, primarily involving epidemics of St. Louis encephalitis.

This inflammatory process involving the central nervous system can result from direct neuron damage by the intracellular replication of the virus or by the immune reaction provoked by the virus (see Chapter 3). The clinical syndromes represent a continuum or spectrum of disease caused by a wide range of viral agents. Such diseases include not only those of the arboviruses but also those associated with childhood infections, herpetic infections, enteroviral infections, influenza, and the like (1–3).

Viral infections of the central nervous system have been classified in terms of the clinical signs and symptoms with which they present themselves (4). According to a commonly used classification, patients can be classified into four groups: aseptic meningitis, encephalitis, paralytic poliomyelitis, and other syndromes. A patient with aseptic meningitis has an acute onset of fever, headache, and symptoms and signs of meningeal inflammation. Cerebrospinal fluid pleocytosis is present; the glucose concentration is normal and routine cultures are negative for bacteria and fungi. The illness has a benign course, and the acute phase usually does not extend more than 10 days.

Patients with encephalitis have fever, headache, and symptoms and signs of meningeal inflammation plus clinical evidence of more deep-seated neurological involvement manifested by one or more of the following: disorientation, excessive drowsiness, stupor, convulsions, tremor, ataxia, focal paralysis, slurred speech, and pathological reflexes. The cerebrospinal fluid findings are similar to those of aseptic meningitis. Encephalitis is a severe illness, and the acute phase and convalescence can have a protracted duration.

The syndrome of paralytic poliomyelitis begins with acute fever, headache, symptoms and signs of meningeal irritation, and persistent flaccid paralysis, without objective evidence of sensory defect or of cortical damage. The cerebrospinal fluid findings mirror those of aseptic meningitis and encephalitis. Other symptoms can be produced,

117

such as febrile headaches, transverse myelitis, and encephalitis involving mainly the brain stem.

In the course of certain viral infections, usually involving children, encephalopathic states may be produced. Encephalopathy implies a significant deep-seated neurological disturbance, including disordered consciousness but not an abnormal number of cells in the cerebrospinal fluid. Because encephalopathic states are closely linked pathogenetically with Reye's syndrome, that entity will not be considered further. This review concentrates on patients having encephalitis, their evaluation, and the practical management of the disease. The diagnosis of encephalitis has ramifications that extend to differential diagnosis, possible pathogenic agents, occurrence of other cases in the community, and the manner in which the patients can be best supported during their illnesses. The last consideration implies an understanding of the disordered physiology of encephalitis. The physician caring for a patient who has encephalitis can best approach the patient's problems by asking, and securing answers to, several critical questions.

QUESTION 1: DOES THE PATIENT REALLY HAVE VIRAL ENCEPHALITIS?

Many clinical conditions mimic viral encephalitis. The physician confronted with the patient who may have viral encephalitis must differentiate the condition from those disorders that are readily treatable. This differentiation is primary and of the utmost importance. The clinical states that have been the most difficult to differentiate from viral encephalitis are granulomatous meningitis (particularly tuberculous meningitis), brain abscess, cortical cerebrovascular accidents with cerebrospinal fluid pleocytosis, and subacute bacterial endocarditis with diffuse cerebral involvement. In some treatment centers, brain tumors, subarachnoid hemorrhages, acute multiple sclerosis, and subdural hematomas have been confused with viral encephalitis. Because many of these conditions are treatable with potent antimicrobial agents or with surgery, physicians caring for a patient who may have viral encephalitis should first consider the treatable conditions that may simulate viral encephalitis.

Case Abstract. A 50-year-old man entered the hospital with a 10-day history of generalized headache, lethargy, and disorientation. His body temperature on admission was 39.1 °C (102.4 °F), his blood pressure was 150/100 mm Hg. The patient muttered constantly and incoherently. His neck was slightly stiff. He had atrial fibrillation with a ventricular response of 70 beats per minute. His white blood cell count (WBC) on admission was 5700, his hemoglobin level was 17.4 gm/dl, and his sodium concentration was 126 mEq/liter. Results of a lumbar puncture were an opening pressure of 320 mm saline, 179 cells, 58% of which were lymphocytes, and a glucose level of 52 mg/dl.

This patient could well have had viral encephalitis. However, the relatively prolonged course before hospital admission is not typical for patients with viral encephalitis, who usually have a short prodromal period. Because the clinical diagnosis could not be made on the initial evaluation of the cerebrospinal fluid, another lumbar puncture was done about 12 hours later. No intravenous glucose was infused for at least 2 hours before the next lumbar puncture to allow for equilibration between the cerebrospinal fluid and plasma glucose levels so that the ratio could be evaluated properly. The next lumbar puncture was essentially similar to the first, with the exception that the cerebrospinal fluid glucose value had dropped to 40 mg/dl, whereas a concomitant plasma glucose value was 150 mg/dl. Further evaluation of this man's condition showed a positive result on the intermediate tuberculin skin test (PPD) and a cerebrospinal fluid protein level that was greater than 100 mg/dl. Later, *Mycobacterium tuberculosis* was grown from the patient's cerebrospinal fluid. Thus, this patient is representative of the situation in which a person presents symptoms as though he might have viral encephalitis but who subsequently is shown to have had initial findings compatible with early tuberculous meningitis. After being placed on antituberculous therapy, the man improved dramatically during the course of several weeks.

The primary concern of the physician faced with a patient presenting with a clinical picture consistent with viral encephalitis is to exclude those entities that may be readily treatable either by antimicrobial agents or by surgery.

QUESTION 2:   DOES THE PATIENT HAVE A COEXISTING DISEASE PROCESS?

Encephalitis may occur as a primary infectious event, but it may also accompany the childhood exanthems such as rubeola, varicella, and rubella. It is frequently seen in the course of mumps. Sometimes it may be associated with respiratory tract infections, particularly influenza, adenovirus, and mycoplasma infections. Encephalitis may complicate herpes zoster infection and be a prominent manifestation of infectious mononucleosis (5). Knowledge of the coexisting disease process may give important clues as to prognosis and may actually affect therapeutic considerations. The following case represents a patient with encephalitis occurring in the course of an influenza A2 (Victoria) infection.

Case Abstract. The patient was a 36-year-old man who was well until one week before admission to the hospital (March 1976). At that time, he developed a "flulike" syndrome with fever, chills, malaise, and cough. Those symptoms continued until the day of his hospital admission, when he became confused, tremulous, and could not recognize members of his family. His temperature at admission was 38.5°C (101.4°F). He was disoriented, unable to respond coherently, and combative when aroused. Myoclonic jerking in his muscle groups was almost continuous, particularly around his shoulders and head. There were no focal neurological signs. Laboratory studies showed a WBC of 6500/mm³ and a hemoglobin value of 14.8 gm/dl. Examination of the cerebrospinal fluid showed 152 white blood cells, 88% lymphocytes, a cerebrospinal fluid/plasma glucose ratio of 64 : 92, and a protein concentration of 80 mg/dl. The electroencephalogram (EEG) showed a diffuse abnormality consistent with encephalitis. The patient continued to be febrile, disoriented, and tremulous until the fifth hospital day, when he became more alert and the focal myoclonic seizure activity disappeared. His fever continued until the seventh hospital day. On discharge, he was alert and oriented and able to ambulate without difficulty. On a follow-up visit three weeks later, he was considered to be neurologically normal. Complement fixation titers for influenza A virus rose from 1 : 16 to 1 : 256 or higher.

The clinical manifestations seen in the course of such a primary disease process may vary widely. Varicella is an example. Mild encephalitis with disorientation and with confusion may progress to semicoma. At times, cerebellar signs predominate. Aseptic meningitis, transverse myelitis, and Guillain-Barré syndrome have been reported related to the occurrence of varicella. About 30% of varicella associated with central nervous system involvement is related to the development of Reye's syndrome. The incidence of encephalitis seen in the course of the common childhood illnesses appears to be decreasing, apparently because of the availability of effective vaccines (rubeola, rubella, mumps).

Being cognizant of the associated disease is important because sufficient knowledge exists of the course of the encephalitic process to guide the physician in formulating a prognosis. In one instance, that guidance may have therapeutic implications. With severe encephalitis associated with infectious mononucleosis, some authorities would recommend a trial of steroid therapy. With the other encephalitides, steroid therapy probably ought not to be used at all or only as an adjunctive maneuver in relieving unrelenting cerebral edema.

QUESTION 3:   DOES THE PATIENT HAVE A FORM OF VIRAL ENCEPHALITIS THAT IS POTENTIALLY TREATABLE, FOR EXAMPLE, HERPES SIMPLEX ENCEPHALITIS?

Herpes simplex virus type 1 (HSV-1) is the most common known agent producing sporadic encephalitis in the Western world today (6–8). The classical configuration of herpes simplex encephalitis is that of an asymmetric encephalitic process simulating an expanding lesion or mass in the temporal lobe. Many of the cases are in mature adults and occur sporadically throughout the year. Features in common with the other encephalitides, such as fever, nuchal rigidity, and disturbances in consciousness, are seen. Herpes simplex encephalitis usually involves primarily the temporal lobes. Therefore, the entrance of the virus into the brain has been postulated to occur along the olfactory bulb and tract. Although primary infection probably accounts for some instances of herpes simplex encephalitis, the majority opinion at present is that most cases probably represent reactivation of a latent infection. Within the temporal lobe, the process is that of a hemorrhagic necrotizing encephalitis with the presence of cells that have Cowdry type A nuclear inclusion bodies. These inclusion bodies on hematoxylin and eosin staining are eosiniphilic and separated from a condensed nuclear membrane by a clear zone. The hemorrhagic necrotic lesion expands and abuts on the subarachnoid space with the consequent leakage of blood. The cerebrospinal fluid findings are those of a mononuclear pleocytosis, generally, but not always, a normal

121

cerebrospinal fluid sugar, and the presence of red blood cells or xanthochromia. The expanding temporal lobe lesion creates a situation in which intracranial pressure tends to increase. Papilledema may be observed. The encephalitic process may extend to adjacent areas of the brain. Characteristic of herpes simplex encephalitis are a disturbance of consciousness and focal neurological abnormalities. A few patients in the prodromal stage may have olfactory hallucinations. On careful neurological examination, most patients with herpes simplex encephalitis have localized neurological abnormalities. The EEG is always abnormal, and it may also indicate a temporal lobe defect. Brain scan, arteriogram, and computerized atrial tomography (CAT) scan also may reveal an abnormality related to the temporal lobe.

Unfortunately, the exact diagnosis of herpes simplex encephalitis today necessitates obtaining tissue from the involved area and then ascertaining the presence of virus by culture, fluorescent antibody studies, and electron microscopy or the presence of intranuclear inclusions by light microscopy. Because the diagnosis of herpes simplex encephalitis implies that a drug may be effective, knowledge of the clinical configuration is mandatory, because the next step is to establish that diagnosis by brain biopsy. Our own practice in evaluating patients with viral encephalitis is to obtain a brain biopsy whenever evidence suggests a process is localized to the temporal lobe or adjacent structures. Unfortunately, no consistently reliable noninvasive methods are now available for establishing the diagnosis of herpes simplex encephalitis early in its course. A fourfold rise in antibody titer occurs late, generally when the disease has already progressed to the point at which the patient may have residual sequelae. Determination of specific antibody in the spinal fluid may suggest the diagnosis, but the antibody may only be present late in the course.

Among the most promising new attempts to diagnose herpes simplex encephalitis by noninvasive procedures is the demonstration of viral antigen by immunofluorescence in cerebrospinal fluid lymphocytes or the demonstration of viral structural and nonstructural proteins by radioimmunoassay or by enzymatic techniques. For example, thymidine kinase, an enzyme coded for by herpes simplex virus, has been found to be elevated in the cerebrospinal fluid of infants having encephalitis due to herpes simplex virus type 2. However, levels of enzymatic activity in infections with HSV-1, the usual agent of adult herpetic encephalitis, are not elevated to the same extent. Because the primary process may be the reactivation of a latent infection, finding the virus in the pharynx is not unusual, but finding it in cerebrospinal fluid is distinctly uncommon.

Faced with the patient who has viral encephalitis and a localizing neurological abnormality on physical examination, one should then attempt to demonstrate that abnormality by EEG, brain and CAT scans, and finally arteriogram. If any of those diagnostic procedures points to asymmetrical involvement of the temporal lobe or its adjacent structures, a brain biopsy is indicated to establish the diagnosis. Once the brain tissue has been obtained, it is submitted for examination by both light and electron microscopy, for immunofluorescent staining, and for culture in an attempt to grow the virus. At present, if the diagnosis of herpes simplex encephalitis is made, the patient may gain access to any of several current experimental clinical studies, from which it appears that adenine arabinoside is the most promising chemotherapeutic agent. IDU (5-iodo-2′ deoxyuridine) is no longer used in the therapy of herpetic encephalitis because of its tendency to cause significant neutropenia and thrombocytopenia at dosage levels that are necessary to control the infection. Intravenous adenine arabinoside is given as a 12-hour infusion for 10 days at a dosage level of 15 mg/kg of body weight per day. Its major side effects are nausea and vomiting. Hematological disturbances are seen with higher dosages, but at the dosage levels mentioned, no significant tendency toward neutropenia or thrombocytopenia has occurred. Carefully controlled double-blind studies now in progress will be necessary to establish the efficacy of adenine arabinoside in herpes simplex encephalitis, but, to date, the results appear to be promising. If left untreated, herpes simplex encephalitis has a case/fatality ratio of approximately 40% and about one half of the survivors have significant sequelae.

QUESTION 4: DOES THE PATIENT HAVE ENCEPHALITIS CAUSED BY AN AGENT THAT HAS PUBLIC HEALTH IMPLICATIONS?

The diagnosis of encephalitis, particularly when caused by an arbovirus, carries with it significant public health ramifications. The occurrence of one or two cases early in the summer may presage the occurrence of epidemic disease. For that reason, cases of encephalitis as clinically defined, whether occurring in children or adults, should be completely investigated from a standpoint of etiology.

Five arboviruses have the potential for causing epidemic central nervous system infection in the United States. Physicians working in a given locality should be aware of the arboviruses that could become active in their community. Eastern equine encephalitis causes a severe encephalitic process with a case/fatality ratio of about 50% and is of particular concern along the eastern and southern Gulf Coasts of the United States. States with particular problems include Massachusetts, New York, New Jersey, Florida, Georgia, and Louisiana. Cases of encephalitis occurring in horses often antedate human cases and may be a signal that human disease may ensue. Western equine encephalitis occurs throughout the western United States, particularly in rural areas, and is a devastating process when it attacks infants. Among infants (< one year) the case/fatality ratio is 50%, and most of the survivors have significant sequelae.

California encephalitis occurs preponderantly in the American Midwest, where it causes an encephalitic illness mostly in children, having a case/fatality ratio of less than 1%. Patients often have rural exposure and are predominantly male. In 1971, Venezuelan equine encephalomyelitis virus crossed the Mexican border into Texas, where it caused human illness and extensive disease in equines. Isolated cases have also been observed in association with an enzootic focus of the virus in southern Florida. Although encephalitis can be seen with this viral infection, the most usual clinical manifestation is that of severe febrile headache.

The arthropod-borne virus that has the greatest public health significance in the United States is St. Louis encephalitis virus (SLEV) (9). Although some years are noted for extensive epidemic activity, 1975 was a particular concern, because SLEV caused 1995 laboratory-documented cases in the United States. Epidemic disease was recorded in Mississippi, Tennessee, and Illinois, with a large epidemic occurring in the southern suburbs of Chicago. SLEV activity extended into New York, New Jersey, and Pennsylvania, as well as the entire American Midwest, South, and Southwest. Since the original epidemic in 1933, other large urban epidemics have occurred, namely in St. Louis in 1937, the lower Rio Grande Valley in 1954, the Tampa Bay area of Florida in 1962, Houston, Texas, in 1964, and Dallas and Corpus Christi, Texas, in 1966. During 1976, SLEV epidemic activity continued to be reported throughout the American Midwest and Southwest.

The epidemiology of St. Louis encephalitis is frequently dictated by the particular mosquito species transmitting the virus. In the midwestern and southern United States, mosquitoes of the *Culex pipiens-quinquefasciatus* complex transmit the disease and are capable of virus transmission within urban areas. In the western United States, where the vector is *Culex tarsalis,* St. Louis encephalitis is closely tied to the transmission of Western equine encephalitis and usually occurs in rural areas. In Florida, another vector (*Culex nigripalpus*) has the capacity for urban transmission. SLEV is a major public health concern in the United States because of its ability to be transmitted extensively within the urban environment. Conditions in cities are conducive to transmission of this virus during the late summer and early autumn because mosquito control programs in most areas often are not comprehensive and such conditions as periodic flooding, substandard housing, and inadequate drainage of surface water exist.

The epidemiology of cases of St. Louis encephalitis in urban areas is distinct. No other viral agent causes the common occurrence of encephalitis in the summertime in urban areas in patients 35 and older. Increasing age-specific incidence and mortality are distinct epidemiologic markers of St. Louis encephalitis in cities. This pattern is sufficiently distinct to enable clinicians or public health epidemiologists who see several cases of encephalitis to make a presumptive diagnosis of St. Louis encephalitis and initiate proper control measures.

The characteristic clinical presentation of St. Louis encephalitis is a symmetrical encephalitic process in older persons. The onset is often abrupt, with fever and headache. Shortly thereafter, patients begin to become disoriented and confused. At that time they are brought to medical attention. On physical examination, patients are febrile, frequently have nuchal rigidity, and usually lie quietly except when aroused. They are disoriented and cannot do simple mental tests. Careful neurological examination may show the presence of tremors, involving the tongue, lips, and mouth. Their hands may move constantly, the tremor becoming intensified at the end of intention. Nystagmus may also be present. If the patients are able to walk, they may be ataxic. Isolated cranial nerve dysfunction, most commonly a seventh nerve paralysis, may be present. Weakness of a particular extremity may be evident at some time in the illness, but signs pointing toward a deep-seated structural lesion usually are absent. Pathological reflexes, such as a positive Hoffmann or a Babinski sign may be found. Frontal lobe release signs are sometimes prominent, with a positive suck and snout reflex as well as positive palmomental reflexes.

Laboratory examination usually reveals leukocytosis, with a predominance of polymorphonuclear leukocytes. Urinalysis may reveal pyuria in the absence of a positive urine culture for bacteria. Papilledema is not present, and the cerebrospinal fluid is under normal pressure. In some patients during the early phase of illness, the spinal fluid may have fewer than 10 lymphocytes per cubic millimeter. If a repeat lumbar puncture is done within 48 hours, however, there generally will be an abnormal number of cells, ranging from 10 to several hundred. Occasionally, early in the course of encephalitis, polymorphonuclear preponderance may be noted, but later, mononuclear cells predominate. The protein is mildly elevated, but the cerebrospinal fluid glucose to plasma ratio always exceeds 0.5. This figure is helpful in differentiating St. Louis encephalitis from other conditions occurring during the summer that may have a bacterial, mycobacterial, or fungal cause.

When confronted with such a patient, the physician must ask the questions noted in this review. Once a diagnosis of encephalitis is established, it should be reported to public health authorities, and then studies should be initiated to document the exact cause. When they are admitted to the hospital, patients with St. Louis encephalitis often are in the primary stages of antibody formation. About 40% of such patients will have a positive HI antibody titer. To establish the diagnosis firmly, a fourfold rise in either the HI or CF antibody titer is necessary. A presumptive diagnosis of SLEV infection may be reached with a serum HI titer of 1 : 80 or a serum CF antibody level of 1 : 16 on a single serum specimen.

Once the diagnosis of St. Louis encephalitis has been established in several persons or the virus has been isolated from mosquitoes, initiation of a surveillance program to look actively for cases is mandated. Mosquito control procedures should be accelerated. Control of the epidemic requires the closest interdigitation between the practicing medical community and public health authorities. In epidemics caused by this virus, the provision to the public of adequate, prompt, and nonsensational information is essential in order to relate the course of the epidemic and the means being taken to stop it. Such information prepares those in the community for the potential necessity to initiate intensified spraying, including the possibility of aerial spraying. The delivery of adequate information is also necessary so that the public is made aware of the conditions that promote mosquito breeding and the necessity for the maintenance of mosquito control programs in the future.

QUESTION 5: HOW CAN THE MOST EFFECTIVE SUPPORTIVE CARE BE DELIVERED TO THE PATIENT WITH ENCEPHALITIS?

Patients with encephalitis are severely ill and need intravenous fluids and, at times, ventilatory assistance. Patients with severe encephalitis may have the syndrome of inappropriate antidiuretic hormone secretion. Physicians must keep this in mind as they plan for fluid replacement. The initial serum sodium concentration may be 130 mEq/liter. There are two explanations for this hyponatremic state: One is volume depletion in the course of fever, sweating, and a relatively increased intake of water by mouth without commensurate increase in sodium chloride intake. Another explanation is the syndrome of inappropriate antidiuretic hormone secretion. In this syndrome, instead of volume depletion the intravascular compartment volume is often normal or increased. Physicians planning intravenous therapy should assume that in the initial phases of the illness, a decreased serum sodium concentration is actually due to volume depletion. Attempts to restore volume by administering normal saline should be instituted. If antidiuretic hormone secretion is inappropriate, however, the hyponatremia will not be corrected, and further administration of sodium chloride in an attempt to correct the hyponatremia may actually injure the patient.

Because patients with both herpes simplex encephalitis and St. Louis encephalitis tend to be elderly and have underlying cardiovascular and renal disease, continued sodium chloride administration may precipitate congestive heart failure. In herpes simplex encephalitis, in which hemorrhagic necrosis and increased intracranial pressure are common, the continued administration of sodium chloride may even aggravate the encephalitic process. In epidemics of St. Louis encephalitis, patients with hyponatremia have died with fluid overload when sodium chloride administration was continued. Since patients with encephalitis are under severe stress, they have attained high steroid outputs in the resting state. Continued high steroid levels may promote a tendency toward salt and water retention and should also be considered in planning intravenous fluid therapy.

Patients with encephalitis are often mildly hypoxic because they fail to ventilate atelectatic pulmonary segments. Oxygen by face mask can relieve this hypoxia, and for the adult patient with underlying myocardial disease, it may be a significant therapeutic adjunct. Alternatively, neurogenic hyperventilation may occur in some patients. In St. Louis encephalitis, although the arterial pCO_2 may thereby be lowered, cerebral blood flow is normal because the disease process has caused loss of the autoregulation of cerebral circulation. The inflammatory process around the respiratory centers may predispose the patient to neurogenic hyperventilation as has been postulated by some observers. During the course of encephalitis, convulsions may occur that must be managed vigorously with anticonvulsant therapy. Some patients die relatively early in the illness because of the severity of the acute encephalitic process. Others succumb because of the complications engendered during protracted convalescence. One must be aware of the tendency toward GI tract ulceration in these latter patients in order to prevent this complication or treat it appropriately when it does occur. Pulmonary emboli may occur and should be looked for with appropriate measures.

Patients with encephalitis are prone to hospital-acquired infections. If an indwelling Foley catheter is inserted into the bladder, it should be attached to a closed-bag drainage system to maintain sterility. Hospital-acquired pneumonia is also common. Establishing the diagnosis of viral encephalitis in such patients is important because its recognition may prevent needless antibiotic therapy that can shape the nature of ensuing secondary bacterial infections. In the early phases of the management of the patients with encephalitis, it is not uncommon to see them treated with multiple antibiotics in the belief that they do have a bacterial infection. The correct diagnosis of encephalitis may obviate the need for such antibiotic therapy.

The physician should also be aware that investigational approaches to the therapy of severe encephalitis do exist, including those caused by HSV and possibly others such as SLEV. In St. Louis encephalitis, there appears to be a relative paucity of interferon in brain in relationship to the amount of virus. The virus, however, when tested in susceptible cell systems, is sensitive to interferon. It may thus be possible in the future to begin controlled studies of interferon induction or administration of exogenous interferon to see whether or not these particular modalities will be of therapeutic value.

CONCLUSION

When confronted with a patient who may have a viral infection of the central nervous system, the physician should classify the patient in terms of the clinical pattern. Patients with clinically defined encephalitis, particularly adults, differ with respect to pathogenic spectrum, differential diagnosis, and prognosis from patients with other central nervous system syndromes (e.g., aseptic meningitis, paralytic poliomyelitis, febrile headache). The physician must ask, and secure answers to, five questions:

1. **Does the patient really have viral encephalitis?** Conditions treatable by antibiotics or surgery should be searched for and carefully excluded.
2. **Does the patient have an unrelated, coexisting disease process?** If the patient has encephalitis in the course of rubeola, this diagnosis carries with it certain implications with regard to prognosis.
3. **Does the patient have a form of encephalitis that is potentially treatable, for example, herpes simplex encephalitis?** Knowledge of the clinical configuration of herpes simplex encephalitis has and will become increasingly important as drugs, such as adenine arabinoside, become available and are shown to be effective in treatment.
4. **Does the patient have encephalitis caused by an agent that has public health implications?** If it can be established that an arbovirus is the cause, potent means exist for control extending to aerial spraying of insecticides to prevent further spread in the community.
5. **How can supportive care be effectively delivered to the patient with encephalitis?** Knowledge of the disordered pathophysiology attendant on severe encephalitis can aid the physician in planning intravenous fluid therapy during the acute stage and in preventing complications that may ensue during protracted convalescence.

DISCUSSION

Question: Shouldn't isolation attempts be initiated for using cerebrospinal fluid and peripheral blood leukocytes, particularly early in the course of the infection, despite the obvious greater yield in brain biopsy?

Dr. Luby: In the infant, such isolation has been successful, but in the adult, unfortunately, it has been rarely successful. The two most promising attempts to diagnose herpes encephalitis noninvasively involve examination of the cerebrospinal fluid. One method that has had indifferent results is to look for herpes antigen in cerebrospinal fluid mononuclear cells; the other is to look at proteins or enzymes coded for by the virus, one of which is thymidine kinase. This method appears to be more promising for HSV-2 in infants than for HSV-1 in adults. At this time, we can't reliably diagnose herpes simplex infection early in its course in adults by noninvasive techniques.

Question: If St. Louis encephalitis is such a problem, how many subclinical cases do we see for every clinical case? Is it possible to develop a vaccine like the TC83, the vaccine that was used to control the Venezuela encephalitis epidemic in Texas?

Dr. Luby: There are probably about 250 infections for every case, and the problem with immunization is that you just don't know which population to immunize. I would never have predicted its northward extension into Illinois or Canada or its eastward extension in 1975. I have no real data that would enable me to decide which population to immunize at the present time.

REFERENCES

1. Illis LS (ed): *Viral Diseases of the Central Nervous System.* Baltimore, Williams & Wilkins Co., 1975.
2. Johnson RT, Mims CA: Pathogenesis of viral infections of the nervous system. *N Engl J Med* 278:23–30, 1968.
3. Johnson RT, Mims CA: Pathogenesis of viral infections of the nervous system. *N Engl J Med* 278:84–92, 1968.
4. Meyer HM, Johnson RT, Crawford IP, et al: Central nervous syn-

dromes of "viral" etiology: A study of 713 cases. *Am J Med* 29:334–347, 1960.

5. Schnell RG, Dyck PJ, Walter EJ, et al: Infectious mononucleosis: Neurologic and EEG findings. *Medicine* 45:51–63, 1966.

6. Leider W, Magoffin RL, Lennette EH, et al: Herpes simplex encephalitis. *N Engl J Med* 273:341–347, 1965.

7. Miller JK, Hesser F, Tompkins VN: Herpes simplex encephalitis: Report of 20 cases. *Ann Intern Med* 64:92–103, 1966.

8. *Neurotropic Viral Diseases Surveillance.* Annual encephalitis summary—1974. Atlanta, Center for Disease Control, US Dept of Health, Education and Welfare.

9. Southern PM, Smith JW, Luby JP, et al: Clinical and laboratory features of epidemic St. Louis encephalitis. *Ann Intern Med* 71:681–690, 1969.

8

Hemorrhagic Fevers

LEON ROSEN

Viruses of various taxonomic groups can produce hemorrhagic phenomena in man. Hemorrhage is common in some human viral infections and rare in others. The term "hemorrhagic fever" has no precise definition, and whether or not a particular viral disease sometimes accompanied by hemorrhage is considered a hemorrhagic fever has been determined by the time and place of its description and the state of contemporary knowledge. For example, if yellow fever were recognized for the first time today, it would be classified as a viral hemorrhagic fever, and it is so listed here. The term "hemorrhagic fever" began to appear commonly in American medical literature about 25 years ago in descriptions of, and discussions on, a severe and often fatal disease of military personnel in the Korean War. Since then, for one reason or another, certain exotic viral diseases have been referred to in the United States (and elsewhere) as hemorrhagic fevers, and these, listed in Table 8.1 in chronologic order of first description, will be discussed here. One should remember that hemorrhage often is not a prominent component of the clinical picture of some of these infections, and as already noted, hemorrhagic syndromes commonly occur in other human viral infections.

All hemorrhagic fevers listed in Table 8.1 are known or suspected to be zoonoses or are transmitted by arthropods or both. They otherwise are diverse with respect to the classification of their viral pathogenic agents (Table 8.2), their epidemiology (Table 8.3), and their geographic distribution (Table 8.4).

Because none of the listed diseases normally occurs in the contiguous United States today, one might ask why they should be considered here. Obviously, with modern air transport, a patient can arrive in the United States within the incubation period of almost any infectious dis-

TABLE 8.1. *Viral Hemorrhagic Fevers by Chronologic Order of First Description*

Yellow fever
Hemorrhagic fever with renal syndrome
  (Korean hemorrhagic fever)
Crimean hemorrhagic fever
Omsk hemorrhagic fever
Dengue hemorrhagic fever
Argentine hemorrhagic fever
Kyasanur forest disease
Bolivian hemorrhagic fever
Marburg disease
Lassa fever

135

TABLE 8.2. *Viral Hemorrhagic Fevers by Classification of Causative Agent*

Flavivirus (group B arthropod-borne virus):
   Yellow fever
   Omsk hemorrhagic fever
   Dengue hemorrhagic fever
   Kyasanur forest disease
Arenavirus:
   Argentine hemorrhagic fever
   Bolivian hemorrhagic fever
   Lassa fever
Bunyavirus:
   Crimean hemorrhagic fever
Unclassified viral agent:
   Marburg disease
Viral cause suspected but agent not yet isolated:
   Hemorrhagic fever with renal syndrome

---

ease acquired almost anywhere abroad. Although that is good reason to become acquainted with the clinical manifestations and epidemiology of exotic diseases, an additional consideration is the propensity for some viral hemorrhagic fevers to be transmitted from person to person in the hospital environment. In view of the very high case/fatality ratio for such hemorrhagic fevers and the absence of readily available means

TABLE 8.3. *Viral Hemorrhagic Fevers by Primary* Method of Transmission to Man*

Tick-borne:
   Crimean hemorrhagic fever
   Omsk hemorrhagic fever
   Kyasanur forest disease
Mosquito-borne:
   Dengue hemorrhagic fever
   Yellow fever
Rodent-associated (arthropods unnecessary):
   Argentine hemorrhagic fever
   Bolivian hemorrhagic fever
   Lassa fever
   Hemorrhagic fever with renal syndrome†
Unknown:
   Marburg disease

---

* Some of the diseases listed can be transmitted secondarily from person to person in a hospital environment without the intervention of an animal host.
† Suspected to be rodent-associated on the basis of epidemiologic data.

TABLE 8.4.   *Geographic Distribution of Viral Hemorrhagic Fevers*

Fever	Location
Yellow fever	African and American tropics
Hemorrhagic fever with renal syndrome	Northern Asia, Europe
Crimean hemorrhagic fever	Central Asia, Europe, ? Africa
Omsk hemorrhagic fever	Central Asia
Dengue hemorrhagic fever	Southeast Asia, Pacific Islands, ? Americas, ? Africa, ? Europe
Argentine hemorrhagic fever	Argentina
Kyasanur forest disease	India
Bolivian hemorrhagic fever	Bolivia
Marburg disease	Africa
Lassa fever	Africa

for active or passive immunization, one can be faced with a situation analogous to caring for a patient having virulent smallpox without the ability to protect attendants by vaccination.

The diseases in which nosocomial spread has been a problem are Crimean hemorrhagic fever, Bolivian hemorrhagic fever, Lassa fever, and Marburg disease. Because almost all the new information on viral hemorrhagic fevers relates to these particular diseases and to dengue hemorrhagic fever, these five diseases will be discussed. For information on other hemorrhagic fevers and for a review of, and references to, older data on the aforementioned five diseases, the reader is referred to previous publications (1-14).

CRIMEAN HEMORRHAGIC FEVER

Crimean hemorrhagic fever was first recognized in 1944 in the Crimea, and its viral pathogenesis was established by the inoculation of human volunteers soon thereafter. Not until 1967, however, was the causative agent propagated easily and consistently in the laboratory (11). With that advance, similar diseases seen in other parts of the USSR (e.g., Central Asian hemorrhagic fever) and Bulgaria were shown to be caused by the same agent (11). Moreover, a virus isolated from humans in Africa about 10 years previously (15) and called Congo virus was found to be antigenically indistinguishable from Crimean hemorrhagic fever virus (16). Whether human infections with Congo virus in Africa are as severe on the average as those with Crimean hemorrhagic fever virus in the USSR remains unknown. Isolates of Crimean hemorrhagic fever/Congo virus have been obtained from humans, *Hyalomma* and other ticks, and cattle, sheep, goats, horses, and other mammals in the USSR, Pakistan, Bulgaria, and Africa (11,17). Serologic studies have extended the known geographic range of the virus to Iran (18) and India (19). Human infection occurs preponderantly among rural populations and apparently results from tick bites, crushing infected ticks with the hand, or exposure to the blood of infected animals. Nosocomial and laboratory infections with the virus are common and apparently result both from direct contact with blood containing virus and from other sources.

The following description of a recent occurrence in one of the major cities in Pakistan (20) illustrates the potential danger of this agent should it be encountered in a hospital environment. A man who lived in a rural area on the outskirts of the city became ill with fever and was admitted to a hospital five days later with massive hematemesis and melena. A laparotomy was performed, and although blood was present in the peritoneal cavity, no perforation or bleeding site was discovered. Four days after surgery, four members of the surgical team became ill, first with fever and later with bleeding. Two of those persons died after a clinical course similar to that of the index case. In addition, at about the same time, a physician who cared for the index case in the hospital ward and the patient's father, who had attended him both at home and on the ward, became ill with the same disease. The father died. Finally, two physicians and one nurse who had attended one of the sick members of the surgical team acquired the disease but survived.

BOLIVIAN HEMORRHAGIC FEVER

Bolivian hemorrhagic fever was first recognized in 1959 in northeastern Bolivia, and its relationship to Argentine hemorrhagic fever became known in 1962. The virus of Bolivian hemorrhagic fever was first isolated the next year (6) and named Machupo virus, after a small river in the endemic area. In 1969, similarities in morphology and morphogenesis between the viral agents of Argentine hemorrhagic fever and Bolivian hemorrhagic fever on one hand and that of the long-known lymphocytic choriomeningitis virus on the other led to the recognition of a new viral family, the arenaviruses. This family also included a number of other viruses that are not known to be pathogenic for man.

Machupo virus has been isolated from the urine of rodents of the genus *Calomys,* and human infection is believed to be acquired from the secretions or excretions of the rodents. An important factor in the epidemiology of Bolivian hemorrhagic fever (and in other diseases caused by arenaviruses) is the ability of the rodent hosts to become chronically infected and shed virus in urine for long periods.

Although person-to-person transmission was not observed in the original outbreak of Bolivian hemorrhagic fever, an especially severe episode of nosocomial spread has been observed after the importation of a case into a nonendemic area of Bolivia (21).

MARBURG DISEASE

Marburg disease was first recognized in 1967 among laboratory workers in Marburg and Frankfurt, Germany, and in Yugoslavia. The workers were exposed to blood, tissues, or cell cultures from monkeys (*Cercopithecus aethiops*) recently imported from East Africa. A very unusual viral agent, morphologically unlike any previously described, was isolated from some of the patients and named Marburg virus.

Secondary person-to-person transmission occurred among medical attendants caring for these patients and, several months later, to the spouse of a primary-case patient. The latter infection is believed to have been transmitted via semen, because Marburg virus has been recovered from semen many weeks after the onset of illness. In all, 29 cases were reported from Germany (with 7 fatalities) and 2 from Yugoslavia. *Cercopithecus* monkeys do not appear to be a reservoir of the infection in nature; experimental infection of those animals is invariably lethal, and no Marburg virus antibody has been found in them in Africa.

The disease or its causative agent was not recognized again until 1975 (22), when a young Australian tourist somehow acquired a fatal infection during extensive travels in Africa and in turn transmitted the disease to his traveling companion and a medical attendant in a hospital. The source of infection for the primary case in this series could not be determined.

By far the most important recognized outbreak of the disease occurred in Africa late in 1976 (23–26). Hundreds of cases of a hemorrhagic disease occurred in southern Sudan and northern Zaire, with a high mortality. Preliminary figures indicate that 59 of 137 cases in Sudan and 325 of 358 cases in Zaire were fatal. The degree of nosocomial spread is indicated by the fact that 30 of the 42 cases in one hospital in Sudan were staff members of the institution. The causative agent of the Zaire outbreak was found to resemble the previous isolates of Marburg virus morphologically but proved to be different antigenically. No information is as yet available on the primary source of human infection in these recent Marburg disease episodes.

LASSA FEVER

Lassa fever was first reported in 1969 when a newly recognized virus was isolated from three missionary nurses who were working in northeastern Nigeria (7). The original patient acquired her infection in Lassa town and was the source of infection for the two other nurses, one of whom was evacuated to New York City. Two laboratory workers in the United States acquired the disease while working with specimens from the original three patients. Two of the latter and one laboratory-acquired infection proved fatal.

In 1970, a second outbreak of the disease, involving at least 28 persons, occurred in Jos, Nigeria, and in 1972, a third, involving 11 persons, occurred in Liberia. Both were nosocomial outbreaks, the virus having been introduced into the hospital environment by a patient with an undiagnosed fever. A fourth outbreak that occurred in Sierra Leone differed from the previous ones in that most patients were infected outside the hospital.

The virus responsible for Lassa fever has been found to be a member of the newly recognized arenavirus group, and an African rodent, *Mastomys natalensis,* has been incriminated as the primary source of human infection.

Lassa fever differs from the two other hemorrhagic fevers caused by arenaviruses in its wider geographic distribution, its higher mortality, and the greater tendency toward nosocomial transmission. In addition to the original case in a patient transported to the United States, patients with the disease have been evacuated subsequently, either knowingly or inadvertently, to London, Hamburg, and Washington, D.C. (27).

DENGUE HEMORRHAGIC FEVER

In 1953, the designation "hemorrhagic fever" was applied to cases of what proved ultimately to be dengue infection in the Philippines. When originally described, Philippine hemorrhagic fever was of unknown cause and its identification resulted from the contemporary interest in the hemorrhagic fever of Korea (hemorrhagic fever with renal syndrome). The first viruses associated with the Philippine hemorrhagic fever were newly recognized serotypes (types 3 and 4) of dengue viruses, and initially the viruses were believed to be particularly virulent agents that often give rise to hemorrhagic infections. Later, when it was found that any of the four dengue serotypes can cause hemorrhagic disease and that such disease occurred widely in Southeast Asia and India, the term "dengue hemorrhagic fever" came into use. In recent years the number of dengue virus infections has increased not only in Southeast and South Asia but also in some Pacific islands (28) and the Caribbean region (29–31). Hemorrhagic manifestations have been associated with that increase.

Dengue hemorrhagic fever differs in a number of important respects from the other hemorrhagic fevers that have been discussed (and the others that have not). First, whether judged by number of cases, number of deaths, or size of population at risk, dengue hemorrhagic fever is more important than all other hemorrhagic fevers combined. Second, unlike all the other viral hemorrhagic fevers considered here, dengue hemorrhagic fever is not restricted to rural areas with particular ecologic conditions. Rather, it is preponderantly an urban disease that can occur in the warmer parts of the world where man creates a suitable environment for its principal vector, the mosquito *Aedes aegypti*. Unlike several of the other viral hemorrhagic fevers, dengue hemorrhagic fever cannot be transmitted directly from person to person in a hospital environment without the intervention of its mosquito vector. Finally, unlike the causative agents of other viral hemorrhagic fevers, dengue viruses and the usual disease caused by them were known long before the hemorrhagic syndrome was described. Thus, one is confronted with the apparent problem of why a "new" disease associated with a well-known infectious agent had suddenly appeared.

Despite considerable differences of opinion about some aspects of dengue hemorrhagic fever (32–34), currently few believe that the disease is "new," although it clearly has been more prevalent in recent years than in the known past. The possible reasons for this increase in prevalence will be discussed below when the pathogenesis of the disease is considered. One disagreement on dengue hemorrhagic fever concerns the exact signs, symptoms, and laboratory findings needed to classify a patient as having the disease. If one accepts the concept that dengue infection associated with hemorrhage and death can be called dengue hemorrhagic fever, then it can be said that the disease was recognized in Philadelphia as early as 1780 (35) and has occurred throughout the world in association with a number of severe dengue outbreaks.

To explain the increase in incidence of life-threatening dengue in recent years, a hypothesis was proposed (32) that holds that severe dengue hemorrhagic fever, and especially the related dengue shock syndrome, is produced by an immunopathologic mechanism elicited by a second, heterotypic, dengue infection occurring during a certain critical period after an initial infection. In this view, the recent increase in incidence of hemorrhagic dengue is the result of the more frequent multiple dengue infections because of the dramatic increase since the end of World War II in the size of urban populations in Southeast Asia, the continued dissemination of the introduced urban vector, *A. aegypti,* and the more frequent movement of virus serotypes and strains from one Asian urban center to another as a result of more frequent travel and more rapid transport of the human population.

Certain investigators had doubts concerning this hypothesis because other viral hemorrhagic fevers also characterized by thrombocytopenia, elevated hematocrit readings, and hypovolemic shock were clearly primary in nature. No data existed that could not be explained by the sequential infection hypothesis, however, until 1972, when an epidemic of dengue associated with severe bleeding, shock, and death occurred on a remote Pacific island. The circumstances of the epidemic appeared to rule out the possibility of secondary dengue infection as a cause of the severe disease observed (36). Later investigations uncovered additional cases of dengue shock syndrome in persons undergoing primary dengue infection (37). Also, studies have recently shown (38) that reduced levels of serum complement observed in dengue shock syndrome that had been hypothesized to occur as a result of antigen-antibody complexes also occur in shock that is clearly nonimmunologic.

In my opinion, explained in detail elsewhere (34), no valid evidence exists to support the concept of secondary infection as the cause of severe dengue hemorrhagic fever. Rather, all the observed data on dengue hemorrhagic fever can be explained by the hypothesis that different strains of dengue viruses of all four serotypes vary in their pathogenic potential and that life-threatening dengue and shock syndrome are relatively rare consequences of dengue infection. In this view, the increased incidence of hemorrhagic manifestations of dengue in recent years is explained by the likelihood that the total number of dengue infections has increased enormously for the reasons mentioned above (i.e., increased urban populations, dissemination of *A. aegypti,* etc.). In other words, the iceberg of total dengue infection has become so large that the visible portion, that is, the severe clinical manifestations, now projects farther out of the water.

CLINICAL OBSERVATIONS

Unfortunately, the clinical manifestations of the viral hemorrhagic fevers are not distinctive enough to enable the clinician to make a differential diagnosis without the use of laboratory tests. Although differences among the various diseases can be demonstrated in a relatively large series of cases, there is enough variability when considering a few patients to render those differences useless. However, a few clinical observations are common to almost all viral hemorrhagic fevers. First, the incidence of hemorrhagic manifestations, despite the names of the diseases, is often low. Second, hemorrhage, when it does occur, almost always begins several days after the onset of fever. Third, leakage of plasma through vessel walls and resultant hypovolemic shock (in the absence of hemorrhage) is seen in many of the diseases. Thrombocytopenia is almost always observed in severe cases. The pathogenesis of plasma leakage and hemorrhage is not known for any of the diseases, but disseminated intravascular coagulation is believed to play a role as the cause of hemorrhage, at least occasionally. Finally, convalescent plasma is believed, but has not been proved, to be useful in treatment of some of the viral hemorrhagic fevers, although not in dengue hemorrhagic fever.

The hazards of attempting to isolate some of the pathogenic agents in the laboratory are considerable, and consequently, isolation procedures can be undertaken only in a very few laboratories in the world.

REFERENCES

1. Gajdusek DC: Acute infectious hemorrhagic fevers and mycotoxicoses in the Union of Soviet Socialist Republics. *Medical Science Publication No. 2,* Washington, DC, Walter Reed Army Medical Center, 1953.
2. Gajdusek DC: Virus hemorrhagic fevers. *J Pediatr* 60:841–857, 1962.
3. Smorodintsev AA, Kazbintsev LI, Chudakov VG: *Virus Hemorrhagic Fevers.* Jerusalem, Israel Program for Scientific Translations, 1964.
4. Casals J, Hoogstraal H, Johnson KM, et al: A current appraisal of hemorrhagic fevers in the USSR. *Am J Trop Med Hyg* 15:751–764, 1966.
5. Casals J, Henderson BE, Hoogstraal H, et al: A review of Soviet viral hemorrhagic fevers, 1969. *J Infect Dis* 122:437–453, 1970.
6. Johnson KM, Halstead SB, Cohen SN: Hemorrhagic fevers of Southeast Asia and South America: A comparative appraisal. *Prog Med Virol* 9: 105–158, 1967.
7. Casals J, Buckley SN: Lassa fever. *Prog Med Virol* 18:111–126, 1974.
8. Work TH: Russian spring-summer virus in India—Kyasanur forest disease. *Prog Med Virol* 1:248–277, 1958.
9. Siegert R: Marburg virus. New York, Springer-Verlag, in Gard S, Hallauer C, Meyer KF (eds), *Virol Monogr* 11:97–153, 1972.
10. International Symposium on Arenaviral Infections of Public Health Importance. *Bull WHO* 52:381–766, 1975.
11. Chumakov MP (ed): Crimean hemorrhagic fever—papers from the Third Regional Workshop at Rostov-on-Don in May 1970. *Misc Pub Entomol Soc Am* 9 (3), August 1974.
12. Mettler NE: Argentine hemorrhagic fever: Current knowledge. *Scientific Publication No. 183.* Washington, DC, Pan American Health Organization, 1969.
13. World Health Organization, Expert Committee on Yellow Fever: Third Report. *Techn Rept Ser No. 479,* 1971.
14. Strode GK (ed): *Yellow Fever.* New York, McGraw-Hill Book Co, 1951.
15. Simpson DIH, Knight EM, Courtois G, et al: Congo virus: A hitherto undescribed virus occurring in Africa: Part 1, Human isolations—clinical notes. *East Afr Med J* 44:87–92, 1967.
16. Casals J: Antigenic similarity between the virus causing Crimean hemorrhagic fever and Congo virus. *Proc Soc Exp Biol Med* 131:233–236, 1969.
17. Berge TO (ed): *International Catalogue of Arboviruses.* Publication No. (CDC) 75–8301. Atlanta, US Department of Health, Education and Welfare, 1975, pp 228, 229.

18. Saidi S, Casals J, Faghih MA: Crimean hemorrhagic fever-Congo (CHF-C) virus antibodies in man, and in domestic and small mammals, in Iran. *Am J Trop Med Hyg* 24:353–357, 1975.

19. Shanmugam D, Smirnova SE, Chumakov MP: Detection of antibodies to CHF-Congo viruses in human and domestic animal blood sera in India. *Trudy Inst Polio Virus Ensef Akad Med Nauk SSSR* 21(2):149–152, 1973. (In Russian)

20. World Health Organization: Viral haemorrhagic fever. *Weekly Epidemiol Rec* 51:261, 1976.

21. Peters CJ, Kuehne RW, Mercado RR, et al: Hemorrhagic fever in Cochabamba, Bolivia, 1971. *Am J Epidemiol* 99:425–433, 1974.

22. Gear JSS, Cassel GA, Gear AJ, et al: Outbreak of Marburg virus disease in Johannesburg. *Br Med J* 11:489–493, 1975.

23. World Health Organization: Viral haemorrhagic fever. *Weekly Epidemiol Rec* 42:327, 1976.

24. World Health Organization: Marburg disease. *Weekly Epidemiol Rec* 43:337, 1976.

25. Update on viral hemorrhagic fever—Africa. *Morbidity & Mortality Weekly Report* 25:339, 1976.

26. Follow-up on viral hemorrhagic fever—Zaire, United Kingdom. *Morbidity & Mortality Weekly Report* 25:378–383, 1976.

27. Follow-up on Lassa fever—Washington, DC. *Morbidity & Mortality Weekly Report* 25:91, 1976.

28. World Health Organization: Dengue fever surveillance in some countries of Asia and the South-West Pacific. *Weekly Epidemiol Rec* 50:269–272, 1975.

29. Dengue—Columbia. *Morbidity & Mortality Weekly Report* 21:187, 1972.

30. Follow-up on dengue—Puerto Rico. *Morbidity & Mortality Weekly Report* 25:7, 1976.

31. Dengue—US Virgin Islands. *Morbidity & Mortality Weekly Report* 26:69, 1977.

32. Halstead SB: Observations related to pathogenesis of dengue hemorrhagic fever: VI. Hypotheses and discussion. *Yale J Biol Med* 42:350–362, 1970.

33. Hammon W McD: Dengue hemorrhagic fever—do we know its cause? *Am J Trop Med Hyg* 22:82–91, 1973.

34. Rosen L: The Emperor's New Clothes revisited, or reflections on the pathogenesis of dengue hemorrhagic fever. *Am J Trop Med Hyg* 26:337–343, 1977.

35. Rush B: An account of the bilious remitting fever, as it appeared in Philadelphia in the summer and autumn of the year 1780, in *Medical Inquiries and Observations*. Philadelphia, Prichard & Hall, 1789, pp 104–121.

36. Barnes WJS, Rosen L: Fatal hemorrhagic disease and shock associated with primary dengue infection on a Pacific island. *Am J Trop Med Hyg* 23:495–506, 1974.

37. Scott RM, Nimmannitya S, Bancroft WH, et al: Shock syndrome in primary dengue infections. *Am J Trop Med Hyg* 25:866–874, 1976.

38. Branson HE, Wyatt LL, Schmer G: Complement consumption in acute disseminated intravascular coagulation without antecedent immunopathology. *Am J Clin Pathol* 66:967–975, 1976.

9

Herpesviruses

JERRY W. SMITH

During the past 25 years, research on herpetic infections of man has increased dramatically. Of the many reasons responsible, two are most prominent: (a) detection of Epstein-Barr virus in lymphoid cells from patients with Burkitt's lymphoma and infectious mononucleosis, and (b) the association of herpes simplex virus type 2 with squamous cell carcinoma of the cervix. Along with these findings, which provided the stimuli for increased research, interest in herpesviruses has been maintained because of the many diseases now known to be caused by these agents.

Of the 25 or so viruses that have been placed in the family Herpetoviridae (herpesviruses) (see Chapter 2), only 4 are known to infect man with any regularity: herpes simplex virus (HSV), Epstein-Barr virus (EBV), cytomegalovirus (CMV), and varicella-zoster virus (VZV). The incidence of infection with each, however, is high, as shown by the fact that persons living past middle age usually have antibody to all 4 viruses.

The herpesviruses have a core of double-stranded DNA (about 8×10^7 daltons) surrounded by a protein coat, which in turn is covered with an outer lipid-containing membrane or envelope. Members of the family are similar morphologically (about 150 nm) but do not share common antigens. The four viruses that infect humans appear to have only minimal immunologic relationships to corresponding viruses of lower animals. Even those that infect humans are not closely related. Of the four, only HSV and VZV have been shown to have antigenic determinants in common, sometimes resulting in cross-reacting antibodies.

The most fascinating feature of herpesviruses, as they relate to human disease, is not their morphological similarities or antigenic relationships to one another, but rather their capacity to establish latent or persistent infections or both, a feature that many investigators regard as the underlying mechanism by which these viruses may be capable of causing malignancies.

HERPES SIMPLEX VIRUS

Herpetic skin eruptions (*herpein;* from Greek, "to creep") have been known since biblical times. As with most viral diseases, however, the most significant advances in our understanding of herpetic infections have taken place in the 20th century. In early laboratory studies, the experimental host range was limited to animals and chicken embryos,

149

systems that, by their nature, limited the number of possible approaches to the problem. The modern age of experimentation really began in the 1950s with the advent of cell culture techniques, which made *in vitro* propagation of the virus possible, thereby adding a new dimension to studies of the relationship of HSV to disease.

One early significant finding, published in the 1960s from laboratories in Germany, England, and the United States, was that HSV could be subdivided into two antigenic types: HSV-1 and HSV-2 (1–4). That finding was followed by the discovery that most oral infections were caused by HSV-1 and most genital infections by HSV-2 (5). With antigenic differences to distinguish the two types, the association of HSV-2 to carcinoma of the cervix became increasingly evident (6,7). The implications of that association, however, have still not been fully clarified.

Diseases of Herpes Simplex Virus

A feature of infections caused by either HSV-1 or HSV-2 is that in many persons the virus, after primary infection, may become latent. With the appropriate stimulus, such as sunlight, fever, menstruation, emotional stress, or various illnesses, the virus may subsequently be reactivated to produce manifest disease. Both primary infections with HSV and endogenous recurrences may be asymptomatic and produce immune responses without overt clinical signs and symptoms.

With the development of laboratory techniques for typing these viruses, one serotype has been found most often associated with herpetic disease in a specific anatomical site. According to the old rule of thumb, HSV isolated from lesions occurring above the waistline is usually type 1, whereas HSV isolated from sites below the waistline is usually type 2 (Table 9.1). There is some evidence to indicate that this pattern may be changing, however, as a result of changes in societal attitudes toward oral-genital contact, so either type of virus may be found in either location (8). Type specificity does not, therefore, seem to be a matter of specific tissue tropism. Furthermore, both types can cause the same spectrum of disease manifestions.

Besides the immunologic differences, the two types can also be distinguished by biologic criteria such as heat stability, the character of the lesions formed on the chorioallantoic membrane of chick embryos,

TABLE 9.1. *Diseases Produced by Herpes Simplex Virus*

Infection	Serotype Usually Isolated	Pattern of Infection	Group(s) Usually Affected
Oral			
Gingivostomatitis	HSV-1	Primary	Children
Pharyngitis, tonsillitis	HSV-1	Primary	Children and adolescents
Herpes labialis		Primary and recurrent*	All ages
Eye			
Keratoconjunctivitis	HSV-1	Primary and recurrent*	All ages
Central nervous system			
Meningoencephalitis	HSV-1, 2	Primary	Neonates (HSV-2), others (HSV-1)
Genital			
Vulvovaginitis, urethritis	HSV-2	Primary and recurrent*	Adolescents and adults
Cervicitis			
Penile lesions, urethritis, prostatitis			
Neonatal	HSV-2	Primary	Neonates†
Skin			
Above waist	HSV-1	Primary and recurrent*	All ages
Below waist	HSV-2		Adolescents and adults
Hands or arms	HSV-1, 2		All ages‡
Traumatic lesions	HSV-1		All ages§
Immune-related			
Eczema herpeticum	HSV-1	Primary	Children
Generalized disease	HSV-1, 2		

* A significant number of infected persons do not develop recurrences.
† Transmitted to the neonate through HSV infection of the birth canal.
‡ All ages (HSV-1); adolescents and adults (HSV-2).
§ Wrestlers (herpes gladiatorum); health personnel exposed to oral secretions and thumb-sucking children (herpetic paronychia).

neurovirulence for mice, and the ability to form plaques in chicken embryo cells.

Table 9.1 lists those diseases known to be caused by HSV and the serotypes usually isolated. Descriptions of the pathogenesis and clinical features have recently been reviewed (7,9–11). With the exception of upper respiratory tract infections, tonsillitis, and pharyngitis, the associations between HSV and most of the clinical manifestations have been known for several years. Because HSV is often shed in the absence of clinical signs or symptoms, it has been difficult to determine whether it is the causative agent of active upper respiratory tract disease. In 1964, Evans and Dick (12) reported that HSV appeared to be associated with tonsillitis and pharyngitis in a group of university students. Only recently, however, has definitive evidence been provided to substantiate those findings and to lead some investigators to the conclusion that respiratory clinical associations represent common forms of primary herpetic infection (13–15).

Because HSV can establish latent infections, the current use of immunosuppressive treatment sometimes leads to reactivation and disseminated disease. Persons whose immune competence is impaired because of malignancies, such as leukemia and lymphoma, often suffer severe recurrences of HSV infections. Primary infection of immunosuppressed patients can likewise have devastating effects. Disseminated disease is common in children having Wiskott-Aldrich syndrome (a sex-linked immunologic disorder that affects both humoral and cellular immunity), in severely malnourished children, or in persons having extensive burns (16,17).

Epidemiology

Both serotypes of HSV are worldwide in distribution. Serologic surveys have shown that neutralizing antibodies are widespread, even in persons living in remote and isolated areas, indicating the universal nature of the infection. However, with either serotype, the frequency of infection in a particular population group appears to vary. The variation in incidence is greatest with HSV-2.

Primary HSV-1 infections are most common in childhood and appear to be acquired through close personal contact with infected persons. The usual manifestation in young children is gingivostomatitis;

in adolescents, it is tonsillitis or pharyngitis. Children from lower socio-economic groups acquire the antibody against HSV-1 earlier than do children from higher socioeconomic groups. The incidence of infection in both groups continues to rise with advancing age, most persons having developed antibodies (predominantly against type 1), by the time they reach middle age. Genital infections, on the other hand, do not occur with any frequency until puberty, when sexual activity intensifies (6,7).

The prevalence of genital infections bears a strong relationship to sexual promiscuity. Prostitutes, for example, show a high incidence of antibodies to HSV-2, which, by contrast, are rarely found in members of religious orders who are bound by rules of chastity (7,18). Many investigators believe that the incidence of genital infections is increasing. However, awareness of such infections has come about only in recent years, with the development of precise and sensitive laboratory techniques for serotyping and an understanding of the venereal nature of HSV-2 transmission. Recent findings suggest that, among venereal diseases, genital HSV is second in frequency only to gonorrhea. Up to 300,000 cases of genital herpes are estimated to occur annually, equivalent to about 10% of the number of cases of gonorrhea (8,19).

Early experiments showed that the presence of neutralizing antibodies to HSV does not protect against exogenous infection. Attempts to immunize persons by injecting autologous virus at sites other than the locus of active infection not only failed to prevent endogenous recurrence but also sometimes resulted in new lesions at the injection site (20). Similarly, autoinoculation can occur in the presence of antibodies, as in the thumb-sucking child who has herpetic paronychia. The latter mode of transmission does not appear to be common, however. Primary infection with HSV generally requires a break in the skin. As age increases, the likelihood of introducing virus by autoinoculation diminishes, and coupled with the presence of neutralizing antibodies in most persons, the possibility of infection at new sites is minimized.

Venereal infections, however, appear to be somewhat different. Persons having humoral and cellular immunity to HSV-1 are capable of contracting infection with HSV-2. Furthermore, this pattern appears to be the preponderant one in lower socioeconomic groups. Although some evidence suggests that primary HSV-2 infection is less severe in those who have immunity to HSV-1, it is clear that genital and non-genital infections can occur in the same person, either at the same time or sequentially (6,7,21). No data are now available on the degree of protection that immunity to HSV-1 provides to a person later exposed to HSV-2 through sexual contact. Apparently, however, any protection that is afforded is only partial.

HSV-2 seems to be transmitted mainly via sexual intercourse. Studies have shown that the risk, for a female, of getting the infection from a male who has a penile lesion is greater than 50% (7). Even with no overt lesions, however, the virus may be present in secretions that are a potential source of infection.

Immunology and HSV Infection

Infection with HSV causes production of antibodies to the virus in the usual sequence. IgM is closely followed by IgA and IgG in time of appearance. Once the infection has been acquired, antibodies, primarily IgG, continue to be present throughout life, and the titers appear to undergo little variation, even with recurrent disease.

Two important roles for HSV antibodies are emerging: (a) neutralization of infectious virus, and (b) complement-mediated lysis of infected cells. All three classes of antibody appear to be capable of neutralizing viral infectivity, whereas only IgG and IgM can mediate cytolysis. Immune destruction of cells occurs because of the presence of specific viral antigens that are inserted into the plasma membrane during the course of HSV infection. Viral antibody binds to these antigens, thereby providing sites for attachment of complement, which results in lysis of the cell (16,21).

In recovery from herpetic infections, the cellular arm of the immune response appears to have the greater role. Children who harbor defects in antibody production (agammaglobulinemia), but whose cellular immune systems are competent, recover from HSV infection. In contrast, those having deficiencies in cellular immunity often have severe disseminated disease.

Most studies on cellular immunity to HSV have been concerned with lymphocyte function. The role of macrophages and polymorphonuclear leukocytes in herpetic infection is still unclear. Lymphocytes separated from the peripheral blood of persons with immunity to HSV have been shown to be cytotoxic for HSV-infected cells, presumably because those cells are specifically reactive with virus-induced surface antigens (7,16). Immune lymphocytes stimulated to blastogenesis by interaction with HSV antigens produce a variety of lymphokines, including migration inhibitory factor and lymphotoxin. Some of the lymphokines, in concert with complement-derived chemotactic factors, attract macrophages that exert toxic effects on both infected and uninfected cells. These cytotoxic effects, combined with interferon (produced by activated immune lymphocytes and infected cells) function to limit the spread of infection and to promote recovery of the affected host (22).

Another mechanism by which lymphocytes may kill infected cells requires HSV antibodies for mediation of cytotoxicity (antibody-dependent lymphocyte cytotoxicity). An interesting difference between this interaction with infected cells and that previously mentioned is that the effector cells may come from either an immune or a nonimmune host. The lymphocytes involved appear to be neither T nor B but rather a nonimmune class called K, or null, cells that act by binding to the Fc portions of HSV immunoglobulins attached to surface antigens in infected cells (16,23).

The complexity of the relationship between host defenses and infected cells appears to be even greater than that already described. Recently, cells infected with HSV have been shown to produce Fc receptors in addition to virus-specific surface antigens (Fab receptors). The possibility therefore exists for an immunoglobulin to bind to the surface of an infected cell by either the Fc (nonspecific) portion or the Fab (specific) portion. The relative importance of these newly discovered "receptors" in herpetic infection remains uncertain but is currently the subject of intense investigation (16,23).

Based on our current understanding, it is clear that resistance to, and recovery from, HSV infections are complex processes that will require many years of investigation. All the known factors of humoral and cellular immunity discovered thus far apparently have important roles in some stages of the disease.

Latency

One feature of latent infection with HSV is that reactivation may occur in the presence of neutralizing antibodies. A recent hypothesis advanced to explain this apparent paradox casts interactions between Fc receptors and antibody to HSV in a central role (24). In latent infection, both Fab and Fc portions of the antibody molecule may be bound to the surface of the infected cell, thus protecting it from destruction by either complement or immune lymphocytes. Reactivation might be expected to occur when those bonds were weakened, resulting in derepression of the viral genome. Viral replication would therefore follow, causing the formation of a lesion.

The anatomical location of latent HSV has long been debated. Recent evidence indicates that sensory ganglia are the source of virus that produces recurrent skin lesions. Early attempts to isolate virus by direct inoculation of ganglionic extracts into cell culture were unsuccessful. Recently, however, herpesviruses have been recovered by cocultivation with susceptible cells from viable semilunar (HSV-1) and sacral (HSV-2) ganglionic tissue obtained at autopsy (10). Demonstration of HSV in ganglia does not preclude the possibility that latent virus may also reside in other kinds of tissue, such as the cervical epithelium. Several animal models are available for studying latent disease, and appropriate techniques have now been developed to facilitate cultivation of ganglionic and malignant tissue. Consequently, further advances in this area should be forthcoming.

Treatment and Control

Over the years, a broad spectrum of treatment modalities, including many folk remedies, have been prescribed in attempts to limit HSV infection or to prevent recurrence. Other than the successful chemotherapeutic regimens for keratoconjunctivitis (see Chapter 16), most therapeutic measures proposed have met with failure. Three general approaches have been used: (a) chemotherapy, (b) agents that inactivate virus, and (c) immunologic approaches, such as vaccination or immune stimulation.

Among the agents known to inactivate HSV *in vitro,* such as ethyl ether, chloroform, and glutaraldehyde, most are applied topically with the aim of eliminating infectious virus directly at the site of the lesion (25,26). The effects of most of these topically applied agents have not been examined under controlled conditions, however, so that their efficacy in activating tissue-bound virus is unknown.

The topical use of dyes to limit infection with HSV warrants special discussion. In 1973, it was reported that photodynamic inactivation of HSV resulted in faster healing of cutaneous lesions and reduction in the frequency of recurrences (26). The procedure, for a time widely used in treatment of herpetic lesions, consisted of topical administration of the photoreactive dye, neutral red, followed by exposure to light. Since then, the dye-light treatment has been steeped in controversy. Li et al. (27) showed that virus that has been inactivated photodynamically can transform cells to oncogenicity in newborn hamsters. This finding has aroused fear that HSV could cause malignancies in man through a similar mechanism. Although the potential threat of the dye-light treatment is still being debated, the controversy may fade for yet another and simpler reason: Recent reports of more extensive trials have shown no therapeutic benefit from the technique (28).

Vaccines made from inactivated preparations of HSV-1 and HSV-2 are now under investigation. It would seem that artificial immunization in persons with recurrent disease may be redundant, however, because these individuals have both humoral and cellular immunity to HSV. Moreover, the response to inactivated virus is predominantly humoral, and it is clear that recurrences take place in the presence of high titers of naturally evoked antibody. To derive benefit from HSV vaccines, children, before the age of primary exposure (less than two years old), would have to be the primary target of immunization with HSV-1. On the other hand, HSV-1 fails to protect against HSV-2 infections. Therefore, immunization of adolescents against this latter virus might pose additional problems that would weigh against its potential efficacy.

Currently, attempts are being made to develop subunit vaccines that are free of nucleic acid in order to eliminate any possibility of vaccine-induced malignancy. Immunizing preparations under consideration are composed mainly of glycoproteins from the outer envelope of the virus. Three major questions concerning these and other HSV vaccines need to be answered before the vaccines can be released for general use: Will they prevent disease? Can they be administered without serious side effects? Are they economically feasible? Several groups are developing subunit vaccines, and some evaluation of their potential benefit should begin to appear in the literature soon.

It should be mentioned that vaccines against recurrent HSV infections have been available for several years in Germany under the names Lupidon H, for HSV-1 infection, and Lupidon G, for HSV-2 infection.

These are heat-killed virus vaccines, containing DNA, and they have not been tested rigorously, either virologically or epidemiologically, under controlled conditions. Any benefit has yet to be proved, although results from some recent trials in Germany are encouraging (29). These killed-whole-virus vaccines are not licensed in the United States, and they probably will not be unless results of well-controlled studies attest to their safety and efficacy.

Another immunologic approach to treatment and prevention of herpetic infections is based on the enhancement of immune defenses with nonspecific stimulators. Trials are currently under way in which BCG, which is used in cancer immunotherapy, and levamisole, an antihelminthic drug, are being evaluated. Regimens based on adoptive transfer of immunity through administration of transfer factor are also under evaluation.

Relationship to Cancer

In the early 1970s, results of independent studies suggested that a link might exist between HSV-2 infection and cervical cancer (6,7,18). That deduction was based on data that showed a higher incidence of neutralizing antibodies to HSV-2 in women with cervical cancer than in matched controls. Those findings were greeted as a breakthrough in some quarters and with skepticism in others. Results of additional studies based on population groups representing different cultures and geographic areas have, for the most part, supported the original observations. Rarely has there been 100% correlation between HSV antibodies and the presence of cervical carcinoma, however. Part of this disparity can be explained by the difficulty in differentiating HSV-2 antibodies from cross-reacting HSV-1 antibodies. Subsequent studies have shown that levels of antibody evoked in response to primary HSV-2 infection are influenced by previous infection with HSV-1, even to the point at which little or no detectable antibody to HSV-2 may appear (21,30,31). This finding is in agreement with the fact mentioned earlier that, in the individual case, primary infection with HSV-1 is frequently followed by infection with HSV-2.

Currently, efforts are being directed at developing more sensitive and discriminating assays for detection of immune responses to HSV-2. One approach that has met with some success is based on the demonstration of antibodies to nonvirion proteins produced early in cellular infection. Several groups, using a variety of methods for preparation of antigens, have shown significant differences between cancer patients and normal control subjects (6,32).

Active herpetic lesions of the cervix are rarely observed in patients who have cervical neoplasia. This finding suggests the possibility that a limited portion of the viral genome may persist in tumor cells that are sufficient to code only for early antigens and not for the full complement of proteins needed to produce infectious virus. In one instance, DNA sequences corresponding to the HSV-2 genome have been reported in an invasive cervical tumor (33). However, confirmation of this observation is still lacking.

The *in vitro* oncogenic potential of HSV has been shown in a series of reports that began in 1971 (34). In these investigations, hamster embryo cells were transformed by virus inactivated with ultraviolet light, and the cells so transformed produced tumors when injected into weanling hamsters. Before their discovery, it had been speculated that HSV might be capable of inducing malignant conversion, largely because of its association with latent infections. Today, both HSV-1 and HSV-2 are being used to transform cells *in vitro* and are proving to be valuable systems for determining the general mechanisms of oncogenesis and the immune response to virus-induced tumors.

Many investigators believe that additional seroepidemiologic studies, such as the kind employed to make the original association between HSV-2 and cervical cancer, may now be of limited use for determining the relationship between HSV and human malignancy. Although most seroepidemiologic studies have shown that cancer patients as a group can be separated from controls, the techniques used in them have proved to be unreliable for separating cases from controls on an individual basis. Accordingly, either new methods will have to be developed or new parameters defined if further diagnostic and epidemiologic advances are to be expected. To overcome some of the deficiencies of past approaches, prospective studies are being developed. One is already underway in Czechoslovakia that potentially may involve 15,000 women (6). In that study, women between the ages of 25 and 45 will be followed up for a total of six years and monitored for evidence of HSV infection and changes in cervical epithelium. It is hoped that such studies will provide the direct evidence needed to determine whether HSV infection precedes the development of malignancy.

159

From the data already accumulated, it is reasonable to consider that if viruses are causative agents of human malignancy, HSV-2 is a strong candidate and probably one of the most feasible to study. The evidence is strong that viral constituents are present in some cervical tumors. A major question concerns the role of HSV in this setting: Is it a causative agent or just a passenger?

Laboratory Diagnosis

Both HSV-1 and HSV-2 can be readily isolated in a variety of cells in culture. The cytopathic effect is characteristic, and a presumptive diagnosis can often be made by experienced personnel within a few days. Making a definitive diagnosis, however, is more difficult. Most procedures for identification of HSV require specific hyperimmune antisera and include neutralization of visible cytopathology, fluorescence microscopy, hemagglutination, and radioisotopic or immunoperoxidase labeling (6,7,16). To determine the antigenic type of a viral isolate, antisera of known specificity are absorbed with heterotypic antigens to remove cross-reactive (common) antibodies. The resulting antisera then contain only residual monotypic antibodies. Seth et al. (35), using absorption procedures, recently demonstrated antigenic differences among HSV-2 isolates from different geographic areas. If antigenic diversity is found to be a constant feature, it is possible that each antigenic type will be further subdivided and arranged into subsets, as has been suggested by those authors.

Absorption procedures can also be used to distinguish antibody responses to HSV-1 from those to HSV-2. McClung et al. (36) described a protocol in which ^{51}Cr release, a complement-dependent assay that measures antibodies to surface antigens of HSV-infected cells, can be used to quantitate both cross-reactive and monotypic HSV antibodies in patient sera. Their procedure should be especially helpful in the analysis of antibody production in persons who have sustained infection with both antigenic types of HSV or in individuals who are subject to frequent recurrences.

From the foregoing discussion, it should be apparent that identification of HSV and of specific antibodies requires the services of an experienced virology laboratory staff. Attempts are being made to create simpler assays. One being developed, for example, utilizes enzymes in a colorimetric test to demonstrate HSV antibodies (37). Such simplified procedures may reduce the requirement for cell culture facilities and allow more laboratories to make routine virological diagnoses.

VARICELLA-ZOSTER VIRUS

VZV is the agent of both varicella (chickenpox), a ubiquitous contagious disease that mainly affects children, and herpes zoster (shingles), a sporadic disease of adults characterized by vesicular eruptions of skin areas defined by dermatomes innervated by specific sensory nerves emanating from dorsal root or cranial ganglia. Research on VZV has generally lagged behind that on HSV. Efforts have been hindered by the lack of a susceptible laboratory animal in which to study the disease process and the peculiarities of the cultured virus. In contrast to HSV, which will replicate to high titer in a variety of cell lines in a few days, VZV replicates only to low titer in cells of primate origin. Moreover, the cultured virus is mainly cell-associated, infection *in vitro* being transferred only to contiguous cells, resulting in a long period (days to weeks) before a diagnosis can be made or for a cell sheet to become completely infected.

Biologically, VZV is similar to HSV in some ways. Both agents produce latent infection of sensory ganglia, from which virus can be reactivated to produce manifest disease, and the viruses are distantly related to one another antigenically. The level of CF antibodies to VZV may rise in persons infected with HSV, and conversely, persons infected with VZV may show a rise in antibody titer to HSV. In most characteristics, however, VZV and HSV are distinguishable. Infection with one does not confer immunity against infection with the other. The patterns of latency and reactivation are dissimilar. The primary disease caused by VZV is varicella. Decades later, the virus may become reactivated to again produce clinical disease. This time, however, the disease is herpes zoster. Recurrences of either manifestation are unusual. This is particularly true concerning second attacks of varicella, as the immune response to the primary infection usually confers full protection. HSV infections, on the other hand, recur frequently, usually at the site(s) of primary infection. Finally, the variety of disease conditions produced by HSV is significantly larger than that produced by VZV (Table 9.1).

161

Isolation of the virus is mainly achieved in cells of human origin. Some of the simian cell lines are susceptible but are less sensitive. The cytopathic effect (CPE) may be slow in developing, sometimes taking up to three weeks. Neutralization tests are not generally used to identify the virus, because they take too long and require virus largely free of host-cell material, a condition that is difficult to meet with VZV.

Most of the procedures previously listed for identification of HSV are also used to identify VZV (38). A commonly used technique is fluorescence microscopy, whereby VZV antigens are detected in cell cultures infected with fluid obtained from active lesions. Identification can often be achieved in about two days. Recent attempts have been made to identify viral antigens directly in skin scrapings or from biopsies of infected skin lesions (38,39). These approaches, which utilize fluorescent antibody or gel precipitation, allow VZV infection to be diagnosed quickly, often in less than 24 hours, and promise to reduce substantially the delays often encountered in identification of infection by this virus.

In the past, CF tests were most often used to demonstrate antibodies to VZV in patients' sera. Because of cross-reactions with HSV, however, this kind of serologic test is being supplanted by the newer and more specific assays such as those described in connection with HSV.

Latency

Infections with VZV, like those caused by HSV, recur in the presence of circulating antibody. Although the site(s) at which the virus persists after an attack of varicella is unknown, it has long been theorized that sensory ganglia are involved. The virus has been demonstrated in affected nerves by electron microscopy and fluorescent antibody procedures. However, most attempts to isolate VZV in cell culture from nervous tissue have usually been unsuccessful (38). A successful isolation from a sensory ganglion has been reported recently by Shibuta et al. (40) using human embryonic lung cells. Confirmation of those results will add strong support to the hypothesis that ganglia harbor latent virus.

Although the details of cell-mediated immunity to VZV are not completely understood, evidence has accumulated suggesting that im-

162

pairment of this function is an important factor in the pathogenesis of herpes zoster. Besides the high incidence of zoster in the elderly, in whom immune capabilities have waned, the use of immunosuppressive drugs increases the likelihood that latent virus may be reactivated and result in severe disease. Malignancies that result naturally from immunosuppression, such as Hodgkin's disease and chronic lymphocytic leukemia, are frequently complicated by herpes zoster.

Children who are immunosuppressed and who come in contact with patients in the active stages of zoster may get varicella, which then becomes disseminated, and immunosuppressed children who have experienced previous VZV infection may develop herpes zoster. Because use of immunosuppressive treatment for malignancies and tissue transplants appears to be increasing, VZV infection will doubtless continue to be a complication of commensurate importance.

Vaccines

As with most viral diseases, vaccination against VZV is viewed by many as the best potential means for preventing serious infection. However, a vaccine made of VZV poses several problems. First, little is known of the mechanism of latency. The use of an attenuated live-virus vaccine, therefore, brings with it the risk of creating a population that might harbor the vaccine virus in a latent state, the consequences of which are unknown except for the possibility that the attenuated virus might be reactivated and lead to unusual forms of herpes zoster or other undesirable side effects. Second, the use of inactivated virus is encumbered with substantial logistical and economic problems. Because the virus replicates in cell culture only to low titer and is cell-associated, the task of developing a concentration of virus great enough to use for immunization and free of cellular contaminants is formidable. Furthermore, the immunity induced by killed-virus vaccines is generally much shorter than that following immunization with live-virus vaccines and must be maintained by repeated injections.

Despite these problems, experimental vaccines have been developed and are undergoing clinical trials. Results from a study in Japan, where an attenuated-live-virus vaccine is being evaluated for prevention of varicella, suggest that the incidence of disease can be considerably reduced (41). However, no VZV vaccine will probably be released for general use until its safety and efficacy are conclusively established.

CYTOMEGALOVIRUS

CMV is morphologically indistinguishable from HSV and VZV and, like them, is capable of establishing long-term latent infections. The virus is highly prevalent throughout the world, only a small percentage of the population apparently escaping some form of infection.

Although CMV can cause severe illness, as with the well-known congenital manifestations of cytomegalic inclusion disease, most infections go unrecognized. Even with cytomegalic inclusion disease, it now appears that many more persons are asymptomatically infected (about 10 : 1) than develop serious sequelae. After infection with CMV, the virus may persist in the blood and be shed for long periods in tears, urine, feces, semen, and secretions from the oropharynx. CMV is also commonly found in cervical secretions, in up to 28% of the patients in some of the groups studied (42), and like other human herpesviruses, may cause infection (usually asymptomatic) in the presence of specific antibody. Table 9.2 gives a summary of the clinical conditions that may accompany infection.

TABLE 9.2. *Diseases Produced by Cytomegalovirus*

Groups	Disease(s)*	Notable Features
Children	Nonspecific, hepatitis, respiratory infection	Prolonged excretion of virus
Adolescents and adults	Nonspecific, mononucleosis,† hepatitis	
Neonates	Cytomegalic inclusion disease	Virus acquired from infected mother
Post-transfusion	Mononucleosis†	Lymphocytes suspected of harboring virus
Post-transplant	Mononucleosis,† pulmonary infection, generalized infection	Immunosuppression enhances infection

* Most infections are inapparent.
† Occurs without heterophile antibodies.

Latency

The mechanism of latency, or persistence, in CVM infections is incompletely understood. Because infections have resulted from blood transfusion and because the virus can sometimes be isolated from the buffy coat of peripheral blood cells, leukocytes have been suspected of having a central role (42,43). Joncas et al. (44), using sensitive DNA-DNA reassociation kinetic analyses, detected the presence of the CMV genome in a cell line derived from the lymphocytes of a congenitally infected infant. In another study, Rapp et al. (45) showed that cells derived from human prostate tissue may also carry the genome. Techniques for detection of integrated CMV genomes by nucleic acid homologies, as in the studies cited, are beginning to be used on a wide scale and promise to clarify many aspects of the mechanism of CMV latency.

Immunosuppression

Patients receiving immunosuppressive therapy because of rheumatic disorders or for renal or bone marrow transplants often develop CMV infections, the incidence sometimes reaching 90% (42,46). In persons who have preexisting antibody to CMV, immunosuppressive therapy probably is the trigger that reactivates endogenous infection. Some investigators have shown that CMV may be transmitted by blood transfusion, a particular risk in patients undergoing open-heart surgery (42,47). Likewise, transfused blood administered to patients receiving renal allographs may be the source of infection, particularly in seronegative persons. Recent studies suggest, however, that the transplanted kidney itself may in some instances be a source of infection (48). Use of nucleic acid hybridization probes to detect latent virus in transplanted tissues should shed light on this problem.

Laboratory Diagnosis

Diagnosis of CMV infection is beset by many of the same problems that are associated with VZV. The virus will replicate only in cells of human origin, and any virus produced generally remains cell-associated (see Chapter 14). The replication cycle in cell culture is slow, and with some virus strains, distinct CPE may not be evident for two weeks or more.

Many techniques are available for identification of CMV and the corresponding antibodies (42). In general, the techniques are similar to those described for HSV and VZV. The most interesting of the newer procedures is concerned with the detection of CMV nucleic acid sequences by hybridization techniques, as mentioned earlier. In addition to the promise the techniques show for demonstrating virus in latently infected cells, they also can be used for analysis of CMV strain variation, an area about which little is known today (49).

Immunization

As with the HSV and VZV, significant problems are also associated with the production of a CMV vaccine. A major area of concern is the malignant potential of the virus. There are substantial gaps in our knowledge of the pathogenesis and mechanism of latency. Apparently the CMV genome can exist in a cell without obvious deleterious effect. When infection is produced, it is inefficient, and large numbers of particles are produced that lack nucleic acid, indicating a defect in viral synthesis. Although these findings suggest some potential for malignancy, the only direct evidence has come from *in vitro* studies in which ultraviolet-irradiated CMV was found to transform cells in culture (34).

Because of the economic and technical problems surrounding immunization with inactivated vaccines, recent efforts have been directed toward developing suitable attenuated-virus vaccines. One vaccine produced by multiple passage of CMV in cell culture (Towne strain) has been undergoing human trials in Switzerland (50). Results show good antibody production, but its protective ability has not yet been determined. Likewise, experimental trials are also underway in the United States using the Towne strain, but as yet no results have been published.

EPSTEIN-BARR VIRUS

EBV, like the other human herpesviruses, is ubiquitous and infects most of the world's population. The virus is also capable of establishing latent infections. Once infected with EBV, a person becomes a lifelong carrier, shedding virus periodically. EBV has been associated with three diseases of man: Burkitt's lymphoma, nasopharyngeal carcinoma, and infectious mononucleosis (see Chapter 10). A causative relationship has not been definitely established with the first two of these diseases. However, the association between EBV and infectious mononucleosis, a disease of children and young adults characterized by fever and enlarged lymph nodes, satisfies most of the criteria needed to confirm EBV as the cause.

Seroepidemiologic studies have revealed that in lower socioeconomic groups and in countries in which lower standards of living prevail, EBV spreads quickly among children, most of whom become subclinically infected before school age (51). Among higher socioeconomic groups, infection with the virus occurs at a later age and is often clinically manifested as infectious mononucleosis. Many of the classic symptoms associated with this form of infection are probably due to the immune response rather than to the direct effects of the virus alone (52). The requirement for infection by EBV is evident, however, as persons who have preexisting antibodies do not develop clinical disease.

The role of the virus is less clear in Burkitt's lymphoma, a malignant lymphoma of children in Africa and New Guinea, or in nasopharyngeal carcinoma, a malignancy common in males of Chinese origin. The viral genome has been detected in cells from patients with either condition, but most of the evidence relating these diseases to human malignancy has come indirectly from seroepidemiologic studies showing that the incidence of antibodies to EBV is higher in patients with either neoplasm than in matched controls. Those who advocate immunization against EBV infection point out that prevention of these malignancies may be the only means of establishing this virus as the causative agent.

Since the discovery by Epstein et al. (53) of a herpeslike virus in cultured cells derived from a Burkitt's lymphoma, research on the virus has been intense. Efforts have recently accelerated with the development of animal models (nonhuman primates) in which EBV disease can be produced experimentally. These developments, along with the many sensitive laboratory techniques that are now available with which to study immunity to EBV and the interaction of EBV with cultured lymphoblastoid cells, suggest that the many questions provoked by the discovery of this virus can now be answered.

REFERENCES

1. Munk K, Donner D: Cytopathischer effect und Plaque Morphologie verschiedener Herpes simplex Virusstamme. *Arch ges Virusforsch* 13:529–540, 1963.
2. Pauls FP, Dowdle WR: A serologic study of *Herpesvirus hominis* strains by microneutralization test. *J Immunol* 98:941–947, 1967.
3. Plummer G: Serological comparison of the herpesviruses. *Br J Exp Pathol* 45:135–141, 1964.
4. Schneweis KE: Serologische untersuchungen zur typendifferenzierung des *Herpesvirus hominis*. *Z Immunitaetsforsch* 124:24–48, 1962.
5. Dowdle WR, Nahmias AJ, Harwell RW, et al: Association of antigenic types of *Herpesvirus hominis* with site of viral recovery. *J Immunol* 99:974–980, 1967.
6. Adam E, Melnick JL: Epidemiological approaches to determining whether herpesvirus is the etiological agent of cervical cancer, in Ito Y (ed): Virus and Cancer, *Prog Exp Tumor Res* in press.
7. Nahmias AJ, Josey WE: Epidemiology of herpes simplex viruses 1 and 2, in Evans AC (ed): *Viral Infections of Humans: Epidemiology and Control*. New York, Plenum, 1976, pp 253–271.
8. Chang T: Genital herpes—another VD on the rise. *Resident and Staff Physician* 20:82–87, 1974.
9. Haynes RE: The spectrum of herpes simplex virus infections in children. *South Med J* 69:1069–1078, 1976.
10. Klein RJ: Pathogenetic mechanisms of recurrent herpes simplex virus infections. *Arch Virol* 51:1–13, 1976.
11. Wheeler CE: Pathogenesis of recurrent herpes simplex infections. *J Invest Dermatol* 65:341–346, 1975.
12. Evans AS, Dick EC: Acute pharyngitis and tonsillitis in University of Wisconsin students. *JAMA* 190:699–708, 1964.
13. Glezen WP, Fernald GW, Lohr JA: Acute respiratory disease of university students with special reference to the etiologic role of herpes-virus hominis. *Am J Epidemiol* 101:111–121, 1975.
14. Fenner F, White DO: *Medical Virology*, ed 2. New York, Academic Press, 1976, p 301.
15. Weathers DR, Griffin JW: Intraoral ulcerations of recurrent herpes simplex and recurrent aphthae: Two distinct clinical entities. *J Am Dent Assoc* 81:81–87, 1970.

16.   Nahmias AJ, Shore SL, Kohl S, et al: Immunology of herpes simplex virus infection: Relevance to herpes simplex virus vaccines and cervical cancer. *Cancer Res* 36:836–844, 1976.

17.   Mintz M: Herpesvirus infections in the postneonatal period, in Drew WL (ed): *Viral Infections: A Clinical Approach*. Philadelphia, F. A. Davis Co, 1976, pp 135–188.

18.   Rawls WE, Kaufman RH: Herpesvirus and other factors related to the genesis of cervical cancer. *Clin Obstet Gynecol* 13:857–871, 1970.

19.   Workshop on the treatment and prevention of herpes simplex virus infections. *J Infect Dis* 127:117–119, 1973.

20.   Goldman L: Reactions of autoinoculation for recurrent herpes simplex. *Arch Dermatol* 84:1025–1026, 1961.

21.   Rawls WE, Adam L, Smith JW, et al: Antibodies to herpesvirus types 1 and 2 and cervical cancer, in *Molecular Studies in Viral Neoplasia*. Baltimore, William and Wilkins, 1975, pp 574–583.

22.   Notkins AL: Viral infections: Mechanisms of immunologic defenses and injury. *Hosp Practice* 9:65–73, 1974.

23.   Russell AS, Kaiser JT: Cell-mediated immunity to herpes simplex virus in man. *J Allergy Clin Immunol* 58:539–545, 1976.

24.   Lehner T, Wilton JMA, Shillitoe EJ: Immunological basis for latency, recurrences, and putative oncogenicity of herpes simplex virus. *Lancet* 2: 60–62, 1975.

25.   Nugent GR, Chou SM: Treatment of labial herpes. *JAMA* 224: 132, 1973.

26.   Felber TD, Smith EB, Knox JM, et al: Photodynamic inactivation of herpes simplex virus: Report of a clinical trial. *JAMA* 223:289–292, 1973.

27.   Li JH, Jerkofsky M, Rapp F: Demonstration of oncogenic potential of mammalian cells transformed by DNA-containing viruses following photodynamic inactivation. *Int J Cancer* 15:190–202, 1975.

28.   Myers MG, Oxman MN, Clark JE, et al: Photodynamic inactivation in recurrent infections with herpes simplex virus. *J Infect Dis* 133:A145–A150, 1976.

29. Nasemann T: Herpes simplex type II vaccine: The favorable European experience. *Int J Dermatol* 15:587–588, 1976.

30.   Smith JW, Adam L, Melnick JL, et al: Use of the ^{51}Cr release test to demonstrate patterns of antibody response in humans to herpesvirus types 1 and 2. *J Immunol* 109:554–564, 1972.

31.   McClung H, Seth P, Rawls WE: Relative concentrations in human sera of antibodies to cross-reacting and specific antigens of herpes simplex virus types 1 and 2. *Am J Epidemiol* 104:192–201, 1976.

32.   Notter MF, Docherty JJ: Comparative diagnostic aspects of herpes simplex virus tumor-associated antigens. *J Natl Cancer Inst* 57:483–488, 1976.

33.   Frenkel N, Roizman B, Cassai E, et al: A DNA fragment of herpes simplex 2 and its transcription in human cervical cancer tissue. *Proc Natl Acad Sci USA* 69:3784–3789, 1972.

34.   Rapp F: Herpesviruses and cancer. *Adv Cancer Res* 19:265–302, 1974.

35.   Seth P, Rawls WE, Duff R, et al: Antigenic differences between isolates of herpesvirus type 2. *Intervirology* 3:1–14, 1974.

36.   McClung H, Seth P, Rawls WE: Quantitation of antibodies to herpes

simplex virus types 1 and 2 by complement-dependent antibody lysis of infected cells. *Am J Epidemiol* 104:181–191, 1976.

37. Gilman SC, Docherty JJ: Herpes simplex virus specific antibodies in human sera detected by the indirect enzyme-linked immunosorbent assay. *J Infect Dis,* in press.

38. Weller TH: Varicella-herpes zoster virus, in Evans AS (ed): *Viral Infections of Humans: Epidemiology and Control.* New York, Plenum, 1976, pp 457–580.

39. Olding-Stenkvist L, Grandien M: Early diagnosis of virus-caused vesicular rashes by immunofluorescence on skin biopsies: I. Varicella-zoster and herpes simplex. *Scand J Infect Dis* 8:27–35, 1976.

40. Shibuta H, Ishikawa T, Hondo R, et al: Varicella virus isolation from spinal ganglion. *Arch ges Virusforchung* 45:382–385, 1974.

41. Asano A, Yazaki T, Miyata T, et al: Application of a live attenuated varicella vaccine to hospitalized children and its productive effect on spread of varicella infection. *Biken J* 18:35–40, 1975.

42. Gold E, Nankervis GA: Cytomegalovirus, in Evans AS (ed): *Viral Infections of Humans: Epidemiology and Control.* New York, Plenum, 1976, pp 143–161.

43. Nankervis GA: Cytomegalovirus infections in the blood recipient. *Yale J Biol Med* 49:13–16, 1976.

44. Joncas JH, Menenezes J, Huang E: Persistence of CMV genome in lymphoid cells after congenital infection. *Nature* 258:432, 1975.

45. Rapp F, Geder L, Murasko D, et al: Long-term persistence of cytomegalovirus genome in cultured human cells of prostatic origin. *J Virol* 16:982–990, 1975.

46. Pagano JS: Infections with cytomegalovirus in bone marrow transplantation: Report of a workshop. *J Infect Dis* 132:114–120, 1975.

47. Armstrong JA, Tarr GC, Youngblood LA, et al: Cytomegalovirus infection in children undergoing open-heart surgery. *Yale J Biol Med* 49:83–92, 1976.

48. Ho M, Dowling JN, Armstrong JA, et al: Factors contributing to the risk of cytomegalovirus infection in patients receiving renal transplants. *Yale J Biol Med* 49:17–26, 1976.

49. Huang ES, Kilpatrick BA, Huang YT, et al: Detection of human cytomegalovirus and analysis of strain variation. *Yale J Biol Med* 49:29–44, 1976.

50. Just M, Buergin-Wolff A, Emoedi G, et al: Immunization trials with live attenuated cytomegalovirus Towne 125. *Infection* 3:111–115, 1975.

51. Evans AS, Neiderman JC: Epstein-Barr virus, in Evans AS (ed): *Viral Infections of Humans: Epidemiology and Control.* New York, Plenum, 1976, pp 209–233.

52. Miller G: Epstein-Barr herpesvirus and infectious mononucleosis. *Prog Med Virol* 20:84–112, 1975.

53. Epstein MA, Achong BG, Barr YM: Virus particles in cultured lymphoblasts from Burkitt's lymphoma. *Lancet* 1:702–703, 1964.

10

Epstein-Barr Virus

I. GEORGE MILLER

Epstein-Barr virus, a lymphotropic herpesvirus, has fascinating biologic properties. Recent experiments in our laboratory have been directed in general at understanding the pathogenesis of human infections associated with the virus. We expect that some of the experiments ultimately will prove useful in understanding the nature of the association between EBV and two human malignancies, Burkitt's lymphoma and nasopharyngeal carcinoma. Also, the experiments may eventually shed light on the way in which the extensive lymphoproliferative responses of infectious mononucleosis are so regularly arrested.

By virtue of its morphological appearance and the structure of its genome, which is linear double-stranded DNA with a molecular weight of about 10^8 daltons, EBV is a herpesvirus. However, unlike the other human herpesviruses, herpes simplex, varicella-zoster virus, and cytomegalovirus, EBV does not have a cytolytic interaction with its host cell but rather a temperate relationship. When EBV is added to cultured human lymphocytic cells or to the lymphocytes of certain nonhuman primates, the cells grow indefinitely thereafter. This process of establishment of permanent cell lines after exposure to EBV is part of the more general phenomenon usually called transformation. There is no proof that formation of cell lines after exposure of human cells to EBV is tantamount to oncogenic transformation. Therefore, I prefer the term "immortalization." In the absence of added EBV, lymphocytes from EBV-seronegative persons do not establish cell lines; instead, after several weeks of abortive replication *in vitro,* which may include rather significant spontaneous DNA synthesis, normal lymphocytes die.

The virus can be titrated by serial dilution and the cultures observed for the appearance of large cell clumps and the outgrowth of lines. Alternatively, it has been shown that the virus may be measured by its ability to stimulate cellular DNA synthesis in exposed cells (1). Serial dilution delays the time at which stimulation of cellular DNA synthesis is evident. In the absence of virus or at dilutions past the "end-point," a transient increase occurs in DNA synthesis that ultimately ceases (Fig. 10.1).

OROPHARYNGEAL EXCRETION OF EPSTEIN-BARR VIRUS WITH IMMORTALIZING PROPERTIES

Virus with the property of immortalizing normal lymphocytes is regularly found in the oropharyngeal secretions of man. This form of the

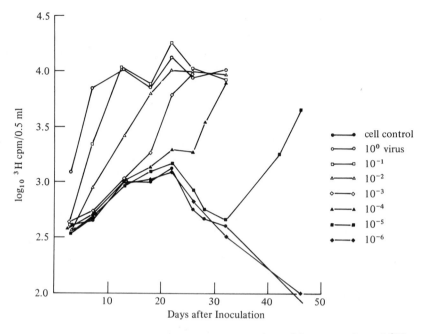

Figure 10.1. Assay for EBV based on stimulation of incorporation of ^{3}H thymidine in human umbilical cord leukocyte cultures. (From ref. 1.)

virus is thought to be transmitted by salivary exchange. I emphasize this point because, with many DNA tumor viruses, transforming particles are relatively rare, often representing a subpopulation that has been damaged or altered in some way so that the particles are no longer lytic. In contrast, excretion of EBV with transforming capacity is a regular event in infectious mononucleosis, and virus excretion persists for many months after the onset of the illness. In a recently completed study, Niederman and co-workers (2) found that the saliva, including that freshly expressed from Stenson's duct, contained the virus long after it was no longer detectable in swabs taken from the tonsils or posterior pharynx (Fig. 10.2). This finding suggests, but does not prove, that some cell in the salivary gland may produce the virus. Exactly which cell in the oropharynx produces EBV is of more than academic interest because the viral genome is found in the epithelial cells of nasopharyngeal carcinoma (3).

Inapparent excretion of transforming EBV is a fairly common event, occurring in about 15% to 20% of healthy persons with antibodies to

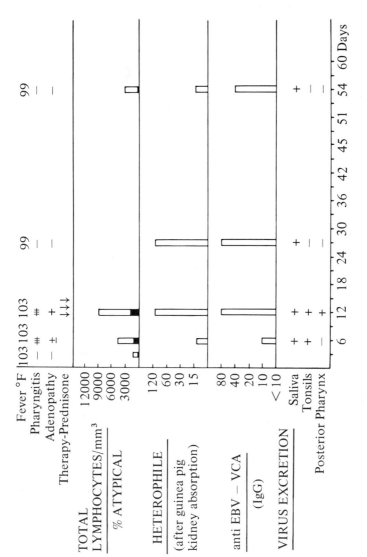

Figure 10.2. Sequence of oropharyngeal and salivary excretion of EBV in a case of infectious mononucleosis studied early after onset. (From ref. 2.)

TABLE 10.1. *Isolations of Leukocyte Immortalization Activity from Throat Washings of 161 Normal Individuals and Patients with Diverse Disease States*

Category	Number Studied	Excretion Rate by EBV Antibody Status		
		Antibody Positive	(%)	Antibody Negative
Healthy	41	5/37	(14)	0/4
Mononucleosis				
Confirmed	27	24/27 ⎫	(78)	
Suspected	13	4/9 ⎬		0/4
Immunosuppressed				
Drugs alone	20	8/15 ⎫	(56)	0/5
Transplants	22	11/19 ⎬		0/3
Chronic renal disease				
Renal failure	26	3/23 ⎫	(19)	0/3
Nephrosis	5	2/3 ⎬		0/2
Other chronic disease				
Nasopharyngeal carcinoma (American)	2	0/2		
Systemic lupus erythematosus	5	2/3		0/2
Total	161	59/138	(43)	0/23

EBV. Virus is not detected in saliva or throat washings of EBV seronegative persons. In certain population groups, particularly those patients having immunosuppressive disease, such as systemic lupus erythematosus or lymphoma, or immunosuppressed renal transplant recipients (4), the frequency of virus excretion rises to 50% or higher (Table 10.1). Thus far, no clinical syndrome has been associated with EBV oropharyngeal excretion in immunosuppressed patients. The supposition is that cells harboring the viral genome persist in the vicinity of the oropharynx and are either activated directly or allowed to be activated through default of immune surveillance mechanisms.

EFFICIENCY OF IMMORTALIZATION

One research goal of some interest is to determine the events required to immortalize a lymphocyte. Immortalization as measured by a microtiter assay follows one-hit kinetics, that is, it takes one biologically active virus particle to immortalize a lymphocyte.

176

The relative proportion of physical virus particles that have transforming capacity is high; we estimate that 1 in 20 to 1 in 50 virions is biologically active. Recently, we developed an assay that we call a "transformed centers" assay for determining the fraction of a virus-exposed lymphocyte population from which cell lines could be established. In a number of experiments we found that 2 to 5 of 1000 mixed human lymphocytes obtained from cord blood established cell lines after exposure to saturating amounts of EBV. When efforts were made to purify the mixed lymphocytes by removing T lymphocytes that are not susceptible to EBV, the transformation rate rose two- to fourfold. These proportions still represent underestimates of the true transformation rate. Even a fully transformed cell line has poor plating efficiency unless accompanied by a system of "feeder cells." When feeder cells are supplied, the plating efficiency rises to 5% or 10%. Thus, if one makes a calculation to correct for plating efficiency, as many as 10% of cells exposed to virus transform (Table 10.2) (5).

This transformation efficiency is higher than that of any other known DNA tumor virus. Paradoxically, this extremely high efficiency of transformation by a ubiquitous virus is not regularly accompanied by oncogenesis. This fact either suggests that immortalization, as measured by formation of a cell line, differs basically from oncogenic transformation or indicates that highly efficient mechanisms exist for detection and elimination of virus-converted cells arising during the course of infection.

Techniques for measuring transformation efficiency will have clinical relevance by making it possible to determine whether population groups differ with respect to the sensitivity of cells to immortalization. We have already found on this basis that adult cells differ considerably from neonatal cells.

TRANSFORMATION OF RESTING LYMPHOCYTES BY EPSTEIN-BARR VIRUS

As noted earlier, lymphocytes differ in their sensitivity to immortalization. This would suggest that in order for EBV to transform them, lymphocytes had to be actively synthesizing DNA at the time of exposure to virus. However, adult blood leukocytes have far fewer cells in DNA synthesis than umbilical cord leukocytes, which makes it appear that

TABLE 10.2. *Efficiency of Transformation of Mixed and T-Cell Depleted Human Umbilical Cord Leukocytes*

Expt.	Feeder Cells	Observed Transformation Efficiency			Observed Plating Efficiency =	Calculated Efficiency of Transformation		
		Mixed Cells	T-Depleted Cells	T-Enriched Cells	÷	Mixed Cells	T-Depleted Cells	T-Enriched Cells
S85	HPC	.001	.004	<.0005	.04	.025	.10	<.01
T26	HUCL	.001*	.004*	ND	.06	.016	.06	ND

* Data for 200 virus-exposed cells/well were used to make this calculation.
HPC = human placental cell (fibroblasts); HUCL = human umbilical cord leukocytes.
T-depleted cells—obtained by centrifugation of cells with sheep erythrocyte rosettes through a Ficoll-Hypaque gradient.

DNA is not required. In fact, the requirement seems to be for quiescent cells not in DNA synthesis. One way this question could be studied *in vitro* would be to analyze the efficiency of EBV immortalization in cells either in active DNA synthesis or in cells that are quiescent. When normal human umbilical cord lymphocytes were examined *in vitro,* about 15% to 20% entered DNA synthesis after one week of culture, less than 1% of the cells having been in DNA synthesis at the time the cultures were initiated. Despite this major difference in DNA synthesis, no major difference in the efficiency of transformation between one-day and seven-day cultures was noted.

This result showed that transformation efficiency was independent of the rate of spontaneous DNA synthesis at the time of exposure to virus. Next, cells in DNA synthesis were eliminated by exposure to 5-bromodeoxyuridine (BrdU) and light and then treated with virus. BrdU and light treatment killed DNA-synthesizing cells but did not alter their susceptibility to stimulation by EBV. However, if BrdU and light treatment were administered after exposure to virus, transformation was reduced. Thus, DNA synthesis was shown to be necessary for transformation, beginning about 24 hours after exposure to virus.

Further corroboration of the finding that EBV transforms a resting lymphocyte was obtained using another technique, namely, velocity sedimentation of the lymphocytes in a gradient of Ficoll-Isopaque, which separates lymphocytes according to size. The large cells in which DNA synthesis is occurring are found in fractions near the bottom of the gradient; a homogeneous population of small, resting lymphocytes is found in the peak fractions of the gradient. The highest efficiencies of transformation were found in the gradient fractions in which cells were not in DNA synthesis.

Taken together these results indicate that EBV immortalizes normal quiescent lymphocytes and that there are no special physiologic requirements before exposure to virus for the event to occur.

A NONIMMORTALIZING STRAIN OF EPSTEIN-BARR VIRUS

Once we had prepared large amounts of immortalizing EBV, we compared the virus with a laboratory strain designated P_3J HR-1, which had been derived from a Burkitt's lymphoma and which had been used as a prototype EBV for many years. To our surprise, we found that current stocks of the P₃J HR-1 virus could not immortalize lymphocytes. Instead, this virus could be measured by its ability to induce "early antigen," representing abortive infection in Raji cells. Conversely, the immortalizing virus could not abortively infect Raji cells. The P₃J HR-1 virus was likewise incapable of stimulating cellular DNA synthesis (6).

This finding has prompted a good deal of controversy in the EBV field. Some investigators believe that the nontransforming abortive replicating virus is the true "wild-type" virus and that transforming variants are mutants or defective viruses. Others, including myself, believe that naturally occurring EBV is an immortalizing virus and that the HR-1 virus is the variant. In favor of the latter position, transforming virus is regularly found in nature, whereas only one example of nontransforming virus has been described.

Recently we began a comparison of the radiobiologic inactivation of the two biotypes of EBV by ultraviolet light and by x-ray. The three functions of the transforming biotype, namely, cell immortalization, stimulation of DNA synthesis, and induction of EBV nuclear antigen, are all highly sensitive to inactivation by ultraviolet irradiation and x-ray. The lethal doses for inactivation of these functions associated with transformation are in the same range as those required for inactivation of plaque formation by HSV-1. This suggests to us that, in some way, a large part of the EBV genome, perhaps all of it, must function to immortalize a cell. We postulate that viral DNA replication is necessary for transformation. By contrast, the function in the nontransforming viral biotype that induces early antigens in nonproducer cell lines containing the viral genome is highly resistant to x-ray and ultraviolet light. This suggests that early antigen induction results from expression of a small part of the superinfecting viral genome and is not the result of viral DNA replication. The resident EBV genome probably plays an important role in early antigen induction. In genome-free lines, early antigen induction is sensitive to radiobiologic inactivation, and the mechanism of antigen induction probably involves viral DNA replication.

The sensitivity of EBV-induced transformation to radiobiologic inactivation stands in contrast to results with other DNA tumor viruses.

For viruses of the papova class, the transforming functions of the virus are more resistant to radiobiologic inactivation than are the lytic functions. For viruses such as herpes simplex, ultraviolet inactivation of the lytic functions of the virus is often necessary to demonstrate transformation.

COMPARING EPSTEIN-BARR VIRUSES FROM THREE ASSOCIATED DISEASES

Having encountered biologic differences between EBVs from Burkitt's lymphoma and infectious mononucleosis, we have been interested to search for evidence on the question whether the great variety of EBV-associated diseases might be the result of serotypic diversity among EBVs. We were able to prepare fairly high-titered EBVs from infectious mononucleosis and Burkitt's lymphoma and have begun to compare them by serum neutralization tests (7). The results to date indicate that cross-reacting antigens exist on the surface of such viruses; but the results also suggest that some unique antigens may be present. Our techniques at best are primitive, and the problem of serotyping herpesviruses is difficult with these methods. So far, not enough strains have been examined to determine whether antigenic variation is epidemiologically significant. Learning whether some virus strains are associated with specific diseases will be important.

Other methods have been used in other laboratories to compare the molecular structure of EBVs. For example, nucleic acid hybridization studies showed that the nontransforming P_3J HR-1 EBV strain contains nucleotide sequences not found in the transforming strains of infectious mononucleosis (8,9). Others have shown that certain cell lines carry the EBV genome as free superhelical circles (10). On the basis of the electron microscopic measurement of such circles, they found that viral DNA associated with lines of Burkitt's lymphoma cells is 5% to 10% longer than circles found in infectious mononucleosis lines.

From this sort of evidence, it is fair to say that the question of the epidemiologic significance of variation among EBVs has not yet been resolved and that more work is needed.

TUMORIGENICITY OF EPSTEIN-BARR VIRUS
IN MARMOSETS

The immortalization phenomenon naturally prompts the question of whether cells immortalized *in vitro* are indeed converted into tumor cells. To answer this question, an animal model was needed. We ultimately selected the cottontop marmoset for the model because its cells could be transformed by the virus, the species was not naturally infected, and previous work by Melendez et al. had shown that the species was susceptible to tumorigenesis by other herpesviruses (11).

In 1 of 4 animals inoculated with autologous EBV-converted cells, malignant lymphoma developed. Five of 16 animals inoculated with cell-free EBV originating from a mononucleosis case developed lymphoma, and 3 other animals developed hyperplasia (12). The lymphoma was histologically of the type formerly classified as reticulum-cell sarcoma but now called immunoblastic lymphoma. Thus, EBV can induce a spectrum of reactions in the cottontop marmoset varying from inapparent infection to malignant lymphoma. Experimental infection of marmosets with EBV mimics many of the features of EBV infection of man. The incubation period of five weeks is similar; generally the same type of antibody responses occur, and virus can be recovered from lymph nodes and blood leukocytes of experimental animals.

Virus is not found in the tumorous lymph nodes, but, in exact analogy with Burkitt's lymphoma, the viral genome is detectable in the tumors by DNA/DNA hybridization. Between 10 and 30 EBV genomes are found for each tumor cell (Fig. 10.3). Once the tumors are placed in culture, the virus is somehow induced and released from the tumor cells.

With any animal model, one must be careful to scrutinize ways in which the model differs from the human counterpart. The marmoset is extremely susceptible to tumorigenesis by EBV. This susceptibility may be related to the route of administration, the immune response of the host, and perhaps the strain of virus used. None of these variables has yet been systematically analyzed. Recently our attention has been drawn to differences on the surface of transformed marmoset cells, and we have pursued the question of surface markers in hopes that it may

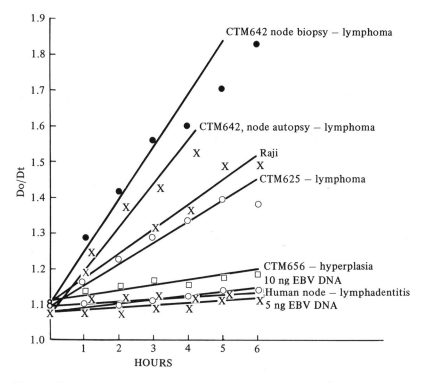

Figure 10.3. Detection of the EBV genome in lymphomas of experiment-ally infected cottontop marmosets. Technique of DNA–DNA association kinetics. (Unpublished data of G. Bornkamm, Erlanger, Germany, and G. Miller.)

shed light on the question of why some immortalizing events are asso-ciated with induction of tumors and others are not.

CELL SURFACE MARKERS ON LYMPHOBLASTOID CELL LINES

Human cell lines transformed by EBV have the surface characteristics of B lymphocytes; that is, they have receptors for C_3b and IgG Fc. James Robinson initially determined whether mononuclear cells of marmoset blood had the same lymphocyte surface markers as human cells. He also analyzed Woolly monkey (WM) lymphocytes. Woolly monkeys were of interest to us because, although their cells may be immortalized *in vitro,* the cells have thus far not proved to be onco-genic when returned to their autologous host. Primary peripheral blood cells from the three species are comparable in the relative proportion of cells with T lymphocyte markers (E), and cells with C' and Fc re-ceptors. However, cell lines transformed *in vitro* are noticeably differ-ent. Whereas the human and WM lines expressed the C' receptor, mar-moset cells did not. WM cells had the receptor for the Fc, but hu-man and marmoset cells did not, at least when an assay based on binding antibody-coated erythrocytes was used. Similar findings were observed when human and marmoset lines transformed by three differ-ent EBV strains were compared. Marmoset lines invariably failed to express the C' receptor, whereas human lines always expressed this receptor.

Those findings could mean that either the virus transforms a differ-ent cell in each species or it transforms the same cell that varies in its expression of surface markers after transformation. To choose be-tween these possibilities, we transformed partially purified subpopu-lations of human and marmoset cells. When preparations of human cells were enriched with respect to cells with the C' receptor, they transformed at an increased rate; conversely, when cells with C' recep-tor cells were removed, transformation efficiency fell. These findings support the idea, first postulated by George Klein of the Karolinska In-stitute in Stockholm, that the C' receptor and EBV receptor are closely linked on the human B lymphocyte (13). Transformation of marmoset cells follows a similar pattern: enrichment with cell preparations with the C' receptor cells results in increased transformation frequencies, and removal of such cells results in a fall in the efficiency of trans-formation.

Several models were initially constructed to explain the findings of different surface markers on cells of different species (Fig. 10.4). The data do not support the idea (Model 2) that EBV transforms a stem cell that later acquires certain lymphocyte surface markers. Nor do they support the model of a different susceptible cell in each species. Instead, the findings support the concept that immortalization is ac-companied by the retention of differentiated surface markers in some instances, such as the C' receptor on human cells and the Fc receptor on WM cells, and is associated with loss of surface markers in others, such as the loss of C' receptor on marmoset cells (14).

These findings suggest that alteration of cell surface properties and

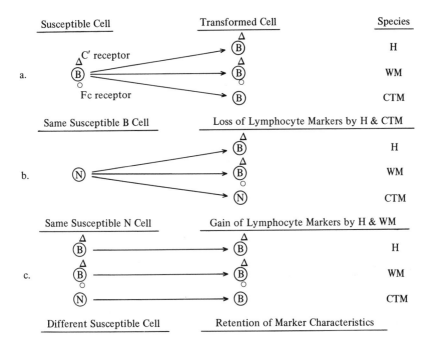

Figure 10.4. Models to account for differences in the expression of lymphoid cell surface markers on human, Woolly monkey, and cottontop marmoset lymphoblastoid cell lines. Available data fit model A, in which CTM cells lose the complement receptor. (From ref. 14.)

alteration of growth properties may be separate events. They lead to the experimentally testable hypothesis that change in cell surface and change in cell growth are both required for oncogenic transformation by the virus. Thus, an important area for future inquiry is a comparison of the surface properties of infectious mononucleosis and Burkitt's lymphoma cells.

SUMMARY

I have tried to indicate some of the subtle biology of EBV in which the problems seem to be of epidemiologic, clinical, and general biological interest. To recapitulate, in the next 10 years we should strive toward having full or even partial answers to some of these questions:

1. What is the basis for the selective interaction between the virus and B lymphocytes?

2. What viral functions (genes) are responsible for immortalization?

3. How are immortalized cells detected and eliminated after mononucleosis?

4. Why does mononucleosis in the older child and young adult differ so much from that in the younger child whose infection is inapparent? Are these age differences related to the differences noted in transformation efficiency and virus production by cells of different ages?

5. Is the same virus associated with infectious mononucleosis, Burkitt's lymphoma, and nasopharyngeal carcinoma?

6. Are nontransforming strains of EBV shed in the oropharynx or anywhere else?

7. Where in the oropharynx does the virus persist? Which cells in the oropharynx produce the virus?

8. What are the crucial cofactors that account for the geographic pathology of Burkitt's lymphoma, with high endemic areas in Uganda and New Guinea?

9. What is the basis for the apparent genetic predisposition of persons of South China origin to nasopharyngeal carcinoma?

10. What measures, in the form of chemotherapy, interruption of transmission, or vaccination, will be applicable to control of infections due to this complex group of viruses?

DISCUSSION

Question: With the evidence of changes in the cell surface during transformation and also the speculation about immunologic surveillance, can these two phenomena be combined in a hypothesis, and can it be said that if something has been deleted, something else has been added to the cell surface? Should one hypothesize that a new antigen is on the surface that might be picked up by immunologic surveillance?

Dr. Miller: To compare the ability of lymphocytes, that are sensitized to EBV-specific antigens to destroy, as target, those cells that do or do not contain a complement receptor but that nonetheless have EBV in them is certainly testable. I, myself, would like to do such experiments or see them done. We don't really know what the function of the complement receptor or the Fc receptor is in the course of immunosurveillance. They may be important, and their absence on the transformed cell may, in fact, prevent that cell from being detected.

These are markers that we have available. We can measure a complement receptor or an Fc receptor, but probably many other markers exist that we don't know how to measure, and many other ligand receptors and surface properties may disappear or reappear in correlation with the things that we can measure. I'm not sure what other surface changes are taking place. I think that the data merely suggest the hypothesis that changes in growth, when accompanied by changes on the cell surface, might have different consequences than just changes in growth in which the differentiated cell surface properties are the same as in the normal cell.

Question: Does the altered B cell in infectious mononucleosis, the cell that's been transformed, produce any type of circulating immunoglobulin? Is IgA antibody to EBV found in the saliva of patients having infectious mononucleosis, that is, the chronic shedder who sheds for weeks or months?

Dr. Miller: I don't know whether immunoglobulin is secreted by the transformed cells or not. Immunoglobulin is on the surface of the transformed cells. What is of interest is that, in infectious mononucleosis, a variety of different immunoglobulins are found on the transformed B cells, suggesting that the disease results from the transformation of several different clones of B cells. In contrast, the immunoglobulin found on lymphocytes in a Burkitt's tumor is usually of one type, that is, monoclonal. This immunologic finding has suggested that one cell gives rise to the Burkitt's lymphoma. The question as to whether the monoclonal immunoglobulin has specificity or whether some of it is directed against the EBV antigen and some of it is heterophile in nature is not resolved, to my knowledge.

I don't know the answer to the second question concerning the presence of IgA. Recently, the Henles reported that in patients with nasopharyngeal carcinoma, the level of serum IgA with EBV specificity increases tremendously. That's about all I know about the IgA.

Question: What is the applicability of the tests you're doing in your laboratory to the early or more definitive diagnosis of infectious mononucleosis?

Dr. Miller: We're interested in being able to make a diagnosis from a peripheral blood smear by counting the cells that actually contain EBV. In other words, our objective would be to make a cytological diagnosis of Epstein-Barr viremia, and that involves knowing which cells are infected and trying to separate and isolate them. Much of the work I described, however, is an effort to determine how the virus actually causes the disease and how it transforms the cell. Another effort has been to determine whether we could identify the virus by looking in the electron microscope at concentrated throat washings. On one occasion, we've succeeded in doing this, but the throat washing was from a patient with a renal transplant who was excreting large quantities of virus. In most patients who have mononucleosis, the amount of virus present in the throat is small, and one cannot identify the virus directly by examining even concentrated saliva.

REFERENCES

1. Robinson J, Miller G: Assay for Epstein-Barr virus based on stimulation of DNA synthesis in mixed leukocytes from human umbilical cord blood. *J Virol* 15:1065–1072, 1975.
2. Niederman J, Miller G, Pearson H, et al: Patterns of excretion of Epstein-Barr virus in saliva and other oropharyngeal sites during infectious mononucleosis. *N Engl J Med* 294:1355–1359, 1976.
3. Wolf H, zurHausen H, Becker V: EB viral genomes in epithelial nasopharyngeal carcinoma cells. *Nature New Biol* 244:245–247, 1973.
4. Strauch B, Siegel N, Andrews L, et al: Oropharyngeal excretion of Epstein-Barr virus by renal transplant recipients and other patients treated with immunosuppressive drugs. *Lancet* 1:234–237, 1974.
5. Henderson E, Miller G, Robinson J, et al: Efficiency of transformation of lymphocytes by Epstein-Barr virus. *Virology* 76:152–163, 1977.
6. Miller G, Robinson J, Heston L, et al: Differences between laboratory strains of Epstein-Barr virus based on immortalization, abortive infection and interference. *Proc Natl Acad Sci* 71:4006–4010, 1974.
7. Miller G, Coope D, Niederman J, et al: Biologic properties and viral surface antigens of Burkitt lymphoma- and mononucleosis-derived strains of Epstein-Barr virus released from transformed marmoset cells. *J Virol* 18:1071–1080, 1976.
8. Pritchett RF, Hayward SD, Kieff ED: DNA of Epstein-Barr virus: 1. Comparative studies of the DNA of Epstein-Barr virus from P_3J HR-1 and B95-8 cells: Size, structure and relatedness. *J Virol* 15:556–569, 1975.
9. Pagano JS, Huang CH, Huang YT: Epstein-Barr virus genome in infectious mononucleosis. *Nature* 263:787–789, 1976.
10. Lindahl T, Adams, A, Bjursell G, et al: Covalently closed circular

duplex DNA of Epstein-Barr virus in a human lymphoid cell line. *J Mol Biol* 102:511–530, 1976.

11.   Melendez LV, Hunt RD, Daniel MD, et al: CEO: Herpesvirus saimiri II: Experimentally induced malignant lymphoma in primates. *Lab Animal Care* 19:378–386, 1969.

12.   Shope T, Dechairo D, Miller G: Malignant lymphoma in cottontop marmosets following inoculation of Epstein-Barr virus. *Proc Natl Acad Sci* 70:2487–2491, 1973.

13.   Yefenof E, Klein G, Jondal M, et al: Surface markers on human B- and T- lymphocytes: IX. Two-color immunofluorescence studies on the association between EBV receptors and complement receptors on the surface of lymphoid cell lines. *Int J Can* 17:693–700, 1976.

14.   Robinson J, Henderson E, Andiman WA, et al: Host determined differences in expression of surface marker characteristics on Epstein-Barr virus-transformed human and simian lymphoblastoid cell lines. *Proc Natl Acad Sci* 74:749–753, 1977.

11

Papovaviruses

GEORGE KHOURY

The discovery of simian virus 40 (SV40) in the early 1960s by Sweet and Hilleman (1) generated great interest and concern among researchers. The virus was present as a previously unknown contaminant in rhesus monkey kidney cultures used in the production of large batches of poliovirus vaccine, and as a result, SV40 was inoculated into thousands of unsuspecting recipients. Additional concern followed from the demonstration by Dr. Bernice Eddy and her colleagues at the National Institutes of Health that the agent produced tumors after inoculation into weanling hamsters (2). Subsequently, a number of laboratories established that SV40 could transform human cells *in vitro* (3–5). The transformation of human-tissue-culture cells by SV40 has been repeated often; to date, SV40 is the only DNA virus shown to be capable of transforming nonlymphoid human tissue. Nevertheless, no human disease state has been linked unequivocally to an SV40 infection. Among the numerous retrospective epidemiologic studies on the human subjects exposed to SV40 through vaccination, only one, by Heinonen et al. (6), has suggested that an increased incidence of malignancies might be related to SV40 inoculation. In that evaluation, the authors concluded that a twofold higher incidence of malignancies, primarily neurological, occurred in the children of mothers who had received inactivated poliovirus vaccine during pregnancy. In the past 10 years, a tremendous effort has been directed at understanding the biologic and biochemical mechanisms underlying the interactions between various types of cells and the two papovaviruses that cause tumors in animals, namely, SV40 and polyoma viruses.

MOLECULAR BIOLOGY

Despite the limited association of SV40 with human neoplasia or other diseases, investigators have been convinced that a thorough understanding of the molecular biology of these small DNA tumor viruses would be crucial to our knowledge of the mechanisms underlying both the normal and the abnormal functions of cells. During the past decade, many advances have occurred in oncogenic virology as a result of developments in the molecular biology of SV40. A partial list of major accomplishments in the field is given here so that the reader may appreciate the progress and current trends.

1. Restriction enzymes were first used to establish a physical map of the SV40 genome (Fig. 11.1) (7,8). The map was fundamental to

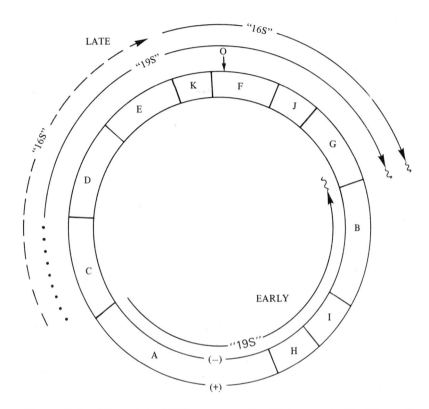

Figure 11.1. Map of the SV40 genome. The 11 fragments (A–K) result-ing from cleavage by restriction endonucleases *Hind* II + III are located with the Endo·RI cleavage site at 12 o'clock. The location and direction of transcription of stable mRNAs (the early 19S and the late 19S and 16S RNAs) are designated by solid arrows. The symbol (→∿) indicates that the stable SV40 mRNAs are polyadenylated at their 3′ termini. (From ref. 95.)

subsequent molecular experiments with SV40 and most other DNA viruses.

2. Viral DNA replication was shown to begin at a unique site (9,10) on the genome and to proceed bi-directionally through a Cairns-type intermediate in a discontinuous fashion on both DNA strands (10–12). The termination of DNA replication is a passive event occurring at a site about 180° from the initiation site (13). After segregation of daughter molecules by an unknown mechanism, a residual gap (14,15) is presumably closed by repair-type enzymes.

3. Transcription of SV40 occurs in two stages (16–18). Before DNA replication, early mRNA is transcribed from one DNA strand (the minus, or E, strand) in a counterclockwise direction probably from a site near the initiation site for DNA replication. After viral DNA replication, abundant quantities of late RNA are transcribed from the other DNA strand (the plus, or L, strand) (19–23). A unit length 26S RNA species is processed first in the nucleus to a 19S molecule and eventually in the cytoplasm to a 16S RNA (24). The 16S late mRNA is translated into the major viral capsid protein (25,26). The important features of the transcriptional system include:

a. a division of early and late mRNA, in terms of the time of transcription relative to DNA replication

b. extensive symmetrical transcription with subsequent processing to remove nontranslated antisense sequences (27,28)

c. control of the steady-state levels of early and late mRNA through control of rates of initiation of transcription (29,30)

d. cytoplasmic processing of a polycistronic message to expose a translation initiation site for the downstream function (i.e., cleavage of 19S to 16S RNA to allow translation of the major capsid protein, VP1) (25,31).

4. Translation of the early viral mRNA accounts for the intranuclear tumor (T) antigen, found both at early times in the lytic cycle and in virus-transformed cells (31,32). Translation of the late viral mRNA results in production of the viral structural proteins, VP1, 2, and 3 (25,26).

5. The SV40 viral DNA was demonstrated to be covalently integrated into the transformed-host-cell DNA (33); the number of copies of this integrated DNA has been quantitated and can involve portions and full copies of the genome (34,35). The site of cleavage for integration within both the viral and host-cell DNA differs from cell line to cell line (36,37). In many of these SV40 transformed lines, virus can be rescued from the covert integrated state by various methods. Transformed cells are characterized by the production of early viral mRNA and T antigen and by the absence of either viral DNA replication or the transcription of late viral mRNA and its translation into late structural viral proteins.

6. The early viral gene product, T antigen, was found to be a DNA-binding protein (38–40), necessary for initiation of viral DNA synthesis, induction of host-cell DNA synthesis, and both establishment

and maintenance of the transformed cells *in vitro* (see below). Because many investigators believe this antigen may be the "transforming protein," a considerable effort is currently being directed toward the elucidation of its structure and mechanism of action.

Despite the vast accumulation of knowledge of the molecular biology of SV40 and polyoma virus, little or no evidence existed before the 1970s that might implicate papoviruses in human disease states. The primary value of these agents was assumed to be in studying control mechanisms for gene expression in mammalian cells. In 1971, however, Padgett and associates (42) isolated a papovavirus, JCV, from the brain of a patient with a degenerative neurological disease, progressive multifocal leukoencephalopathy. The viral cause for the disease was suggested earlier by ZuRhein and Chou (43), who demonstrated the presence of papovalike viruses in the nuclei of oligodendrogliocytes in areas of involved brain tissue. Although the viral agent cultivated by Padgett had the morphological characteristics of a papovavirus, its ability to grow only on primary human fetal glial cells and the lack of structural antigenic relationship to SV40 and polyoma viruses suggested that JCV was a new human papovavirus.

Shortly thereafter, Weiner and colleagues (44) isolated and characterized papoviruses from the brains of two patients with PML. Both of these agents cross-reacted strongly in all immunologic tests with SV40 and were subsequently shown by cleavage with restriction endonucleases and nucleic acid hybridization tests to be strains of SV40 (45,46). Subsequent isolations and immunologic identification of viral agents from the brains of patients with PML all showed papoviruses of the JCV-type, strongly associating JCV virus with PML (47,48).

Another human papovavirus, BKV, was isolated in 1971 by Gardner et al. (49) from the urine of a patient who had undergone immunosuppression and renal transplantation. BKV could be propagated in human cells and caused hemagglutination of human type O erythrocytes. Whereas, clear-cut cross-reactivity occurred among the T antigens (the early virus-specific and probably virus-coded intranuclear antigens) induced by SV40, JCV, and BKV, respectively, the late viral structural or coat proteins of these three agents had only weak antigenic similarities (42,49–52).

Viruses immunologically and biologically indistinct from BKV have been repeatedly isolated from the urine of renal transplant patients (51,53,54). Takemoto et al. (55) isolated this virus from the urine and brain tumor (reticulum-cell sarcoma) of a patient with the Wiskott-Aldrich syndrome.

The isolation of these viruses solely from humans and their preferential growth in human cells in tissue culture suggest that BKV and JCV are human papovaviruses. However, the best evidence that JCV and BKV are human viruses is derived from studies of antibody levels in the human population. Antibodies to SV40 are uncommon in humans (56) (perhaps 3% of the general population), whereas antibodies to both BKV and JCV are prevalent in a variety of diverse human populations. Shah and colleagues (57) found that the prevalence of BKV antibodies detected by hemagglutination inhibition (HI) rose from 50% to 100% in the sera of Maryland residents between the ages of 3 and 11 and then declined to 67% in the group 35 years and older. Similar results were found by Mantyjarvi et al. (58) for a Finnish population and by Gardner (59) for an English population. A similar prevalence of antibodies (69%) against JCV detected by HI was reported by Padgett and Walker (60) in sera of a randomly selected population of residents in Wisconsin that included all age groups. The highest rate of conversion to seropositivity occurred during the first 14 years of life, but in no group were antibodies found to be universally present. Although these data clearly indicate that BKV and JCV are human viruses that induce antibodies early in life, at present no clinical syndrome is associated with seroconversion. It is possible either that human papovaviruses may remain latent and become activated later in life or that reinfection of immunologically incompetent persons may occur. These alternatives will be considered below.

ANTIGENIC PROPERTIES OF THE VIRUS

During its lytic cycle in African green monkey kidney cells, SV40 induces formation of at least three early antigens before the replication of viral DNA. The best characterized of these is the T, or tumor, antigen, first detected by Black and his colleagues (61) using an indirect immunofluorescent test with the serum from an SV40-tumor-bearing hamster. More recently, it has been shown that the papovaviral T an-

197

tigens are virus-coded proteins and gene products of the early region of the genome. Considerable debate continues about the size of T antigen; but the best estimates from SDS-polyacrylamide gel electrophoresis would suggest that it has a molecular weight between 70,000 and 100,000 daltons. Although the function of T antigen is not known, studies with temperature-sensitive mutants of SV40 indicate that the early function is required to (a) start new rounds of viral DNA synthesis, (b) transform cells, and (c) maintain certain characteristics of the transformed phenotype (62–69).

T antigen has been shown to be a DNA-binding protein with preferential affinity for certain regions of the SV40 genome (38–40). One can speculate that it functions as an enzyme, possibly as an endonuclease, a polymerase, or a polymerase modifier. Because T antigen is expressed in transformed cells, it may function by stimulating new rounds of DNA replication in cells that would normally remain in a resting state (70). A knowledge of the function of T antigen would therefore seem to be crucial to the understanding of the function of the papovavirus genome in both lytically infected and transformed cells.

The human papovaviruses similarly induce the synthesis of a T antigen early in the lytic cycle and in transformed cells. The BKV and JCV T antigens cross-react strongly (as determined by various immunologic assays) with the SV40 T antigen as well as with each other. In an attempt to compare directly the T antigens of BKV with those of SV40, analysis of the tryptic peptides of the immunoprecipitated antigens of these viruses has been undertaken by two groups of investigators. The suspicion that these proteins were similar was confirmed by the finding that the major portions of the corresponding tryptic peptides were identical (K. Rundell, P. Tegtmeyer, and G. diMayorca; D. Simmons and R.G. Martin, personal communications). Although the precise function of T antigen remains unclear, the similarities of the T antigens from the various papovaviruses will be discussed below.

Tumor-specific transplantation antigens (TSTA) of papovaviruses have been localized on the surface of transformed tumor cells. They can generally be demonstrated by an *in vivo* assay of immunity to tumor transplantation. Animals immunized by an injection of virus, or by injection of irradiated or heterologous transformed cells, became resistant to a subsequent challenge by cells transformed by the same papovavirus which were tumorigenic for nonimmunized animals. Some evidence exists that TSTA is also present on membranes of lytically infected cells (71). Furthermore, Anderson and colleagues (72,73) recently showed that a preparation of partially purified T antigen can induce immunity to tumor transplantation, suggesting that TSTA may be related to T antigen. Indeed, it may be that only one early polypeptide is specified by the SV40 genome and that the early antigens, T and TSTA, may be different peptide determinants on the same protein. A limited body of information suggests that the TSTAs of the various papovaviruses differ. D. Mason and K. K. Takemoto (personal communication) found that immunization of hamsters with SV40 virus, although sufficient to protect animals against tumorigenesis by SV40, was insufficient to protect against transplantation with cells of tumors induced by BKV or JCV. More recently, Padgett and collaborators (74) immunized weanling hamsters with JCV, BKV, or SV40 and challenged each group of animals with JCV or SV40 tumor cells. Both JCV-immune and SV40-immune animals showed resistance to challenge with homologous, but not with heterologous, tumor cells. BKV-immunized animals were susceptible to both types of tumor cells.

The late antigens (structural or V) can be detected late in the lytic cycle and are absent in transformed cells in which only the early events of papovaviral-cell interactions are expressed. Whereas complement fixation and immunofluorescence studies reveal few, if any, cross-reactions among the JCV-BKV-SV40 viruses (49,51,52), a definite relationship has been established by immune electron microscopy (49,50, 75), and by viral neutralization (51) and hemagglutination inhibition tests (49).

COMPLEMENTATION STUDIES

The ability of various papovaviruses to complement or provide functions necessary for the growth of another virus has been shown in several studies. Mason and Takemoto (76) first described a situation in which "early" SV40 mutants (tsA mutants) that are incapable of growth at 40°C, because of a block in DNA synthesis, could produce plaques in monkey cells that were simultaneously infected with BKV. Subsequently, C.J. Lai, P. Howley, and G.C. Fareed (personal communication) independently showed that the presence of a coinfecting BKV genome allows the SV40 tsA DNA to replicate efficiently at the nonpermissive temperature, suggesting that the BKV early function, expressed as T antigen, is capable of supporting SV40 DNA synthesis. A. J. Levine and colleagues (personal communication) showed, in separate studies, that SV40 coinfection of monkey kidney cells will permit the replication of an early (DNA minus) adenovirus 5 (Ad5) mutant at the nonpermissive temperature in these cells. In an analogous fashion, BKV will permit the DNA replication of this same Ad5 tsA mutant in human cells at the nonpermissive temperature. BKV, however, does not seem to complement late temperature-sensitive mutants of SV40 at the nonpermissive temperature (76). Although the molecular basis of this defect is not known, C.J. Lai et al. (unpublished data) believe phenotypic mixing of defective SV40 and competent BKV capsomeres may occur and result in a defective particle.

THE PAPOVAVIRAL GENOME

The papovaviruses are 45 nm particles with a characteristic icosahedral symmetry and a molecular weight of about 3×10^7 daltons. Within the protein capsid resides a supercoiled double-stranded DNA of molecular weight 3.6×10^6 daltons associated with histones derived from the host cell and perhaps the viral structural protein, VP3. The DNAs of BKV and JCV are slightly smaller than those of SV40 (77,78). On the basis of specific digestion with restriction endonucleases, each has been shown to be unique (Table 11.1). The SV40 genome has been extensively mapped. A limited cleavage map of BKV has been established by Howley and his co-workers (Fig. 11.2) (79). The considerable defectiveness that occurs within the JCV isolate and the difficulty associated with growing and plaquing this agent have thus far prevented the mapping of its genome.

Of particular interest to us was the determination of the polynucleo-

TABLE 11.1. *Restriction Endonuclease Cleavage Sites*

Enzyme	SV40	BKV	JCV
R·*Eco* RI	1	1	1
R·*Hpa* I	3	0	1
R·*Hpa* II	1	0	–
R·*Bam*	1	1	–
R·*Hind* II	5	0	–
R·*Hind* III	6	4	3
R·*Hae* II	1	0	–

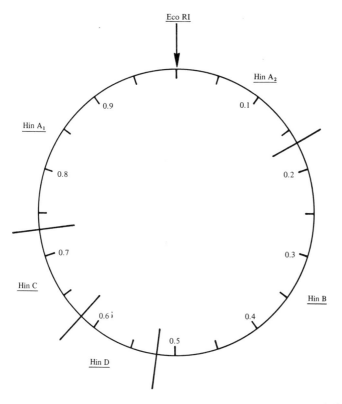

Figure 11.2. The *Hind* III restriction enzyme cleavage map of the BKV genome. (From ref. 79.)

tide sequence homology among SV40 and the other papovavirus genomes. Using a number of different hybridization techniques that differ in their degree of stringency, we were able to show that between 10% and 20% of the SV40 and BKV genomes showed close homology (Fig. 11.3) (77), and up to 50% showed a weaker homology (Fig. 11.4) (80). These regions of strong and weak homology, respectively, were mapped to specific regions of the DNA. The maximal homology was located in various segments of the late region of the genome, whereas little or no homology was found in the early region. This result was particularly surprising, because the T antigens, which exhibit strong cross-reactivity by immunologic assays, are encoded by the early DNA in contrast to the late structural antigens (VP1 and VP2), which show weak immune cross-reactivity between various papoviruses and are

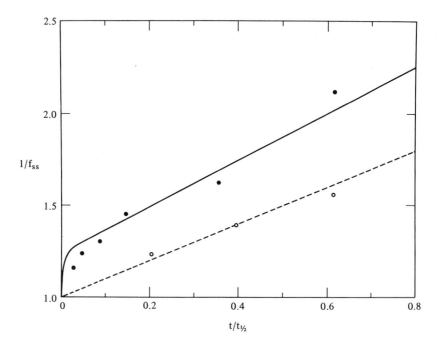

Figure 11.3.   Reassociation of radiolabeled BKV DNA in the presence of an excess of unlabeled SV40 DNA. ^{32}P-labeled BKV DNA (10^5 counts/min/μg) was allowed to reassociate at a concentration of 1.18×10^{-2} μg/ml in the presence ($\bullet$) or absence ($\circ$) of a 100-fold excess of un-labeled, fragmented SV40 DNA in 0.14 M sodium phosphate buffer at 60°C. The percent DNA in duplex molecules was assayed by hydroxyapatite chromatography. The $t_{1/2}$ for the reassociation of ^{32}P-labeled BKV DNA alone (17.1 hr) was the average of six separate determinations. The dashed line is the theoretical curve for the reassociation of BKV DNA alone; the solid line represents the theoretical curve for the renaturation of labeled BKV DNA in the presence of a 100-fold excess of SV40, assuming 20% polynucleotide sequence homology. (From ref. 77.)

encoded by the late region of the genome. Recent studies have pro-vided explanations for these apparently paradoxical situations.

One speculation to explain the apparent lack of homology in the early regions of BKV and SV40 is that the hybridization techniques used could not show the homology as a result of the high degree of mis-match in the DNAs. Using the technique developed by J. Ferguson and R. Davis, N. Newell et al. (personal communication) cleaved both SV40 and BKV DNAs with Eco.R1 endonuclease and ligated the cleaved linear molecules. Covalently linked heteroduplex dimers were

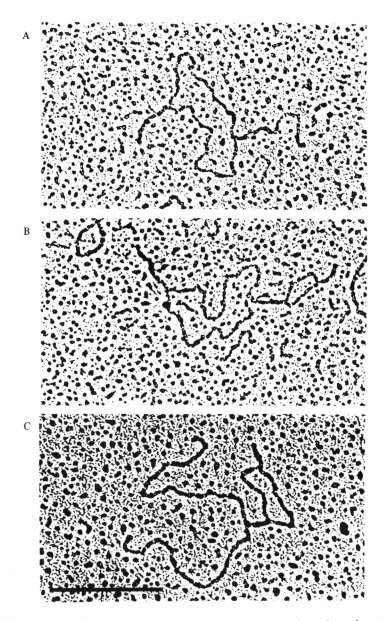

Figure 11.4. Heteroduplex analysis of the sequence homology between the DNAs of SV40 and BKV. Electron micrographs of SV40-BKV heteroduplexes mounted by the formamide isodenaturing technique (41) and rotary-shadowed with platinum-palladium. Panels show heteroduplex molecules spread in formamide concentrations of (A) 40%, (B) 50%, and (C) 60%. Bar represents 1 μm. (From ref. 80.)

denatured, allowed to "snap back," and examined at low formamide concentrations. Under these conditions, homologous areas of considerable mismatch will snap back. About 80% of the BKV and SV40 DNAs form a duplex hybrid including most of the early region. This could explain the cross-reaction of the T antigens by immunofluorescence and the similarity in tryptic peptides. Under these nonstringent conditions, the remaining 20% heterology is concentrated at the regions associated with initiation and termination for SV40 DNA replication.

The second paradox, namely, the strong homology in certain segments of the late region contrasted with the very weak immunologic cross-reaction between SV40 and BKV V-antigens (capsid proteins), appears to have been solved recently by Shah and co-workers (81). Those investigators have shown that if intact virions are used to produce anti-V serum, little cross-reactivity between papovaviruses can be detected. If SDS-treated virions are used as immunogens, however, antisera with considerable cross-reactivity among all the papovaviruses are found. Thus the homologous DNA segments in the late region of the papovavirus genomes seem to code for a portion of the structural polypeptides that is hidden within the virion and is not exposed on the surface during infection. SDS treatment, however, destroys the integrity of the capsid and renders this peptide segment immunogenic.

DNA homology studies have recently been extended to include JCV (82,83) and have established that homology is detectable by hybridization techniques between the genomes of JCV and BKV and between the genomes of JCV and SV40. By analysis of the kinetics of DNA-DNA reassociation, Howley et al. (77) detected 25% polynucleotide sequence homology between JCV and SV40 DNAs (Fig. 11.5). This method of analysis detects 20% polynucleotide sequence homology between the genomes of BKV and SV40. Further studies showed that the homologous sequences shared between JCV and SV40 DNAs are a subset of the homologous sequences detectable between JCV and BKV, establishing that common polynucleotide sequences are shared by all three of these primate papovaviruses (82,83).

ONCOGENICITY AND TRANSFORMATION STUDIES

The potential malignancy of the two human papovaviruses, JCV and BKV, has been established by the demonstration of their ability to in-

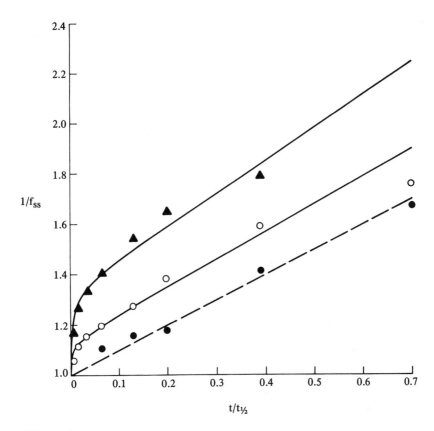

Figure 11.5. Kinetics of reassociation of radiolabeled JCV DNA alone and in the presence of unlabeled SV40 DNA or unlabeled BKV DNA. Mechanically fragmented, heat-denatured, ^{32}P-labeled JCV DNA (sp. act. 2.7 $\times$ 10^5 cpm/μg) was allowed to reassociate at a concentration of 7.29 $\times$ 10^{-3} μg/ml alone ($\bullet$), in the presence of a 391-fold molar excess of fragmented unlabeled SV40 DNA (O) or in the presence of a 370-fold molar excess of fragmented unlabeled BKV DNA ($\blacktriangle$) for varying periods in 0.14 M sodium phosphate buffer (pH 6.8) at 60°C. The fraction of radiolabeled DNA remaining single-stranded (f_{ss}) was assayed by hydroxyapatite column chromatography. The $t_{1/2}$ for the ^{32}P-labeled JCV DNA reassociating alone (30.6 hr) was the average of six determinations. The dashed line is the theoretical curve of the second-order reaction of the reassociation of ^{32}P-labeled JCV DNA alone. The lower solid line is the theoretical curve for the renaturation of labeled JCV DNA in the presence of a 391-fold molar excess of unlabeled SV40 DNA, assuming 11% polynucleotide sequence homology. The upper solid line is the theoretical curve for the renaturation of labeled JCV DNA in the presence of a 370-fold molar excess of unlabeled BKV, assuming 25% polynucleotide sequence homology. (From ref. 82.)

duce tumors in newborn hamsters and to malignantly transform animal cells in culture. Compared with JCV, BKV is weakly oncogenic when inoculated into newborn hamsters (84). JCV is highly oncogenic when inoculated intracerebrally (85). Although no reports have been published of transformation of cells in culture by JCV, a number of laboratories have shown that BKV can transform hamster cells (86–88) and primary nude mouse kidney cells (89). In addition, Takemoto and Martin (90) showed that purified BKV DNA is capable of transforming hamster embryonic kidney cells.

The criteria for the viral specificity of transformation with papovaviruses have been (a) the demonstration of the viral-specific (T) antigen, (b) the rescue of virus from transformed cells when cocultivated with permissive cells, and (c) the demonstration of viral polynucleotide sequences in the transformed cell genome. P.M. Howley and M.A. Martin (manuscript in preparation) detected viral sequences in three lines of hamster embryonic kidney cells that had been transformed with BKV DNA. Not only were those cell lines positive for T antigen but also intact virus was rescued from the population of transformed cells from which they had been cloned. Approximately 2.7, 3.1, and 5.3 copies of the viral genome were present, respectively, per diploid eukaryotic genome in each of the three cell lines examined. Furthermore, using each of the four fragments derived by treating BKV DNA with R. *Hind* restriction endonuclease as probes, Howley and Martin showed that the BKV genome was uniformly represented in the cell line containing 5.3 copies per diploid genome.

POTENTIAL RELATIONSHIP OF PAPOVAVIRUSES TO HUMAN DISEASE

The single most striking biologic characteristic of papovaviruses is their ability to transform cells *in vitro*. Therefore, it is not surprising that much effort has been directed toward attempting to associate papovaviruses with human cancers. A number of studies have suggested that papovalike particles can be detected in certain human tumors by either electron microscopy or by immunoreactivity with SV40 T antisera. In one set of experiments, Weiss et al. (91) found an antigen in three of eight meningiomas that was reactive with antiserum to SV40 antigens. In addition, two of those cell lines became positive for

V antigen, and viruslike particles were visible by electron microscopy after fusion with monkey kidney cells, suggesting that the papovaviral genome had been induced to replicate. In contrast, B. Hirt and his colleagues (personal communication) were unable to detect SV40-related T antigen in any of 26 meningiomas they examined.

In another study, SV40 T antigen and viral particles were found in association with metastases in a patient with malignant melanoma (92). Those cells, however, could not be propagated, so no further experiments could be undertaken to confirm these findings.

The isolation of human papovaviruses has provided a new impetus to search for these agents in human tumors. At present, a number of screening studies are in progress that include attempts to detect papovavirus T antigens in cell nuclei, studies to find antibodies to papovavirus in the sera of cancer patients, and experiments designed to establish the presence of viral sequences in the DNA extracted from tumors or from tumor-derived cell lines. Preliminary evidence shows that nucleic acid hybridization studies with human papovavirus DNA may be fruitful (93). However, establishing the significance of any results will require considerable work and a number of control studies.

In recent studies, J. Costa and A.S. Rabson (personal communication) were unable to detect BKV T antigens in BKV-transformed cells examined by immunofluorescence with sera from cancer patients. The association of JCV with the degenerative chronic neurological disease PML, however, seems firmly established. Papovavirus particles have been detected in the brains of almost every patient with this disease and have been specifically localized in areas of demyelination. Viral agents isolated from these lesions have been papovaviruses of the JCV-type in all but two cases.

More recently, a degenerative neurological disease has been described in macaques that, pathologically, is identical to human PML, providing an animal model for the disease (94). Papovavirus particles are associated with areas of demyelination in the brains of affected monkeys as consistently as found in the human disease. Virus was isolated from two cases of PML in monkeys, and in each case the isolate was identified as SV40 by antigenic analysis and characterization of the genome by restriction enconucleases (94). SV40 is ubiquitous in the monkey colony in which the simian disease has been identified. Likewise, there is serologic evidence in humans for a high incidence of exposure to JCV. Each of these viruses in its natural host thus seems to be associated with a well-defined disease, which, although rarely manifested, is an expression of the same underlying pathological process. In both man and monkey, the disease occurs almost exclusively in association with chronic infectious or malignant diseases, either or both of which may signal impairment of the immune system. Accordingly, some degree of immunologic insufficiency may also be required before PML becomes manifest. Considerable effort is currently being directed toward testing this hypothesis by attempts at isolating additional human papovaviruses in relation to specific human diseases. The technology already developed for analyzing virus cell interactions with SV40 and polyomavirus provides the basis for many of these studies.

DISCUSSION

Question: Do animals with SV40-associated tumors have antibodies against the virus, and if not, why not?

Dr. Khoury: Animals with SV40-associated tumors have antibodies at least to the tumor antigen (T antigen), which is the internal antigen coded for by the early region of the genome. The reason they have antibodies to T antigen is that all tumor or transformed cells, not only infected cells, contain the T antigen in their nuclei. As the tumor becomes necrotic and breaks down, the antigen is expressed as an immunogen, and the animal makes antibody against it. The chances are that the animal has little or no antibody response against V or capsid antigen because, in the tumor cell, the information that is coded for by the late region of the genome (i.e., capsid protein) is not expressed because the virus is not replicating.

Question: What evidence exists that SV40 is the only virus able to transform human cells, and what about EBV and herpesvirus transformation effects?

Dr. Khoury: I think there probably is evidence for EBV transformation of human cells. To my knowledge, there isn't any evidence for

This chapter was written with the assistance of Peter Howley, Mark Israel, Diane Solomon, and Ching-Juh Lai of the National Institutes of Health.
Portions of this chapter have previously been published in *Origins of Human Cancer,* Cold Spring Harbor Laboratory, New York, 1977.

herpes simplex transformation of human cells. SV40 clearly has been shown to be capable of transforming human cells. One unpublished report has suggested that transformation of human cells can be induced by JCV; but that finding certainly must be further substantiated.

Dr. Miller: This word "transformation" means many different things. One question is this: Is there any evidence that SV40 transforms human cells into tumor cells?

Dr. Khoury: In studies done in the early 1960s, terminal cancer patients underwent skin biopsies, fibroblasts from which were transformed *in vitro* by SV40 and subsequently injected back into the patients. After about six weeks, each of these patients developed a nodule that subsequently totally regressed. Thus, even in the terminal cancer patient, the SV40-transformed cells were rejected, probably because of the antigenicity conferred by SV40 to the surface of the transformed cells.

   More recent substantiation of an immunologic barrier to transplantation of SV40-transformed human cells does come from studies with nude (athymic) mice. Although there are some contradictory reports, Dr. Carlo Croce and his colleagues at the Wistar Institute have recently obtained rather convincing data that indicate the SV40-transformed human cells can produce tumors in these animals.

REFERENCES

1.   Sweet BH, Hilleman MR: The vacuolating virus, SV40. *Proc Soc Exp Biol Med* 105:420–427, 1960.
2.   Eddy BE, Borman GS, Grubbs GE, et al: Identification of the oncogenic substance in rhesus monkey kidney cell culture as simian virus 40. *Virology* 17:65–75, 1962.
3.   Shein HM, Enders JF: Transformation induced by simian virus 40 in human renal cell cultures: Morphology and growth characteristics. *Proc Natl Acad Sci* 48:1164–1172, 1962.
4.   Koprowski H, Ponten JA, Jensen R, et al: Transformation of cultures of human tissues infected with simian virus SV40. *J Cell Comp Physiol* 59: 281–292, 1962.
5.   Ashkenazi A, Melnick AL: Tumorigenicity of simian papovavirus SV40 and of virus-transformed cells. *J Natl Cancer Inst* 30:1227–1265, 1963.
6.   Heinonen OP, Shapiro S, Monson SC, et al: Immunization during pregnancy against poliomyelitis and influenza in relation to childhood malignancy. *Int J Epidemiol* 2:229–235, 1973.
7.   Danna K, Nathans D: Specific cleavage of simian virus 40 DNA by re-

striction endonuclease of Hemophilus influenzae. *Proc Natl Acad Sci* 68: 2917, 1971.

8. Danna KJ, Sack GH, Nathans D: Studies of simian virus DNA: VII. A cleavage map of the SV40 genome. *J Mol Biol* 78:363–376, 1973.

9. Danna KJ, Nathans D: Bidirectional replication of simian virus 40 DNA. *Proc Natl Acad Sci* 69:3097–3100, 1972.

10. Fareed GC, Garon CF, Salzman NP: Origin and direction of simian virus 40 deoxyribonucleic acid replication. *J Virol* 10:484–491, 1972.

11. Nathans D, Danna KJ: Specific origin in SV40 DNA replication. *Nature (New Biol)* 236:200–202, 1972.

12. Fareed GC, Khoury G, Salzman NP: Self-annealing of 4 S strands from replicating simian virus 40 DNA. *J Mol Biol* 77:457–462, 1973.

13. Lai CJ, Nathans D: Deletion mutants of simian virus 40 generated by enzymatic excision of DNA segments from the viral genome. *J Mol Biol* 89: 179–193, 1974.

14. Laipus PJ, Sen A, Levine AJ, et al: DNA replication in SV40 infected cells: X. The structure of the 16 S gap circle intermediate in SV40 DNA synthesis. *Virology* 68:115–123, 1975.

15. Chen MCY, Birkenmeier E, Salzman NP: Simian virus 40 DNA replication: Characterization of gaps in the termination region. *J Virol* 17:614–621, 1976.

16. Aloni Y, Winocour E, Sachs L: Characterization of the simian virus 40-specific RNA in virus-yielding and transformed cells. *J Mol Biol* 31: 415–463, 1968.

17. Oda K, Dulbecco R: Induction of cellular mRNA synthesis in BSC-1 cells infected by SV40. *Virology* 35:439–444, 1968.

18. Sauer G, Kidwai G: The transcription of the SV40 genome in productively infected and transformed cells. *Proc Natl Acad Sci* 61:1256–1263, 1968.

19. Lindstrom DM, Dulbecco R: Strand orientation of simian virus 40 transcription in productively infected cells. *Proc Natl Acad Sci* 69:1517–1520, 1972.

20. Khoury G, Byrne JC, Martin MA: Patterns of simian virus 40 DNA transcription after acute infection of permissive and nonpermissive cells. *Proc Natl Acad Sci* 69:1925–1928, 1972.

21. Sambrook J, Sharp PA, Keller W: Transcription of simian virus 40: I. Separation of the strands of SV40 DNA and hybridization of the separated strands to RNA extracted from lytically infected and transformed cells. *J Mol Biol* 70:57–71, 1972.

22. Khoury G, Martin MA, Lee TNH, et al: A map of simian virus 40 transcription sites expressed in productively infected cells. *J Mol Biol* 78: 377–389, 1973.

23. Sambrook J, Sugden B, Keller W, et al: Transcription of simian virus 40: III. Mapping of "early" and "late" species of RNA. *Proc Natl Acad Sci* 70:3711–3715, 1973.

24. Aloni Y, Shani M, Reuveni Y: RNAs of simian virus 40 in productively infected monkey cells: Kinetics of formation and decay in enucleated cells. *Proc Natl Acad Sci* 72:2587–2591, 1975.

25. Prives CL, Aviv H, Gilboa M, et al: The cell-free translation of SV40 messenger RNA. *Cold Spring Harbor Symp Quant Biol* 39:306, 1974.

26. Roberts BE, Gorecki M, Mulligan RC, et al: Simian virus 40 DNA

directs synthesis of authentic viral polypetides in a linked transcription-translation cell-free system. *Proc Natl Acad Sci* 72:1922–1926, 1975.

27.   Aloni Y: Extensive symmetrical transcription of simian virus 40 DNA in virus-yielding cells. *Proc Natl Acad Sci* 69:2024–2029, 1972.

28.   Khoury G, Howley D, Nathans D, et al: Post-transcriptional selection of simian virus 40 specific RNA. *J Virol* 15:433–437, 1975.

29.   Laub O, Aloni Y: Transcription of simian virus 40: Regulation of simian virus 40 gene expression. *J Virol* 16:1171–1183, 1975.

30.   Gilboa E, Aviv H: Preferential synthesis of viral late RNA by nuclei isolated from SV40 lytically infected cells. *Cell* 7:567–573, 1976.

31.   Prives C, Aviv H, Gilboa E, et al: The cell-free translation of early and late classes of SV40 messenger RNA, in Haenni AL, Beaud G (eds): *In vitro* transcription and translation of viral genomes, pp 305–312. *Inserm* QW 160–135. *Les Colloques de Inserm* 47:305, 1975.

32.   Ahmad-zadeh C, Allet B, Greenblatt J, et al: Two forms of simian-virus-40-specific T-antigen in abortive and lytic infection. *Proc Natl Acad Sci* 73:1097–1101, 1976.

33.   Sambrook JF, Westphal H, Srinivasan PR, et al: The integrated state of viral DNA in SV40-transformed cells. *Proc Natl Acad Sci* 60: 1288–1295, 1968.

34.   Gelb LD, Kohne DE, Martin MA: Quantitation of simian virus 40 sequences in African green monkey, mouse and virus-transformed cell genomes. *J Mol Biol* 57:129–145, 1971.

35.   Botchan M, Ozanne B, Sugden B, et al: Viral DNA in transformed cells. III. The amounts of different regions of the SV40 genome present in a line of transformed mouse cells. *Proc Natl Acad Sci* 71:4183–4187, 1974.

36.   Ketner G, Kelly TJ, Jr: Integrated simian virus 40 sequences in transformed cell DNA: Analysis using restriction endonucleases. *Proc Natl Acad Sci* 73:1102–1106, 1976.

37.   Botchan M, Topp W, Sambrook J: The arrangement of simian virus 40 sequences—the DNA of transformed cells. *Cell* 9:269–287, 1976.

38.   Carroll RB, Hager L, Dulbecco R: Simian virus 40 T antigen binds to DNA. *Proc Natl Acad Sci* 71:3754–3757, 1974.

39.   Tenen DG, Baygell P, Livingston D: Thermolabile T (tumor) antigen from cells transformed by a temperature-sensitive mutant of simian virus 40. *Proc Natl Acad Sci* 72:4351–4355, 1975.

40.   Jessel D, Landau T, Hudson J, et al: Identification of regions of the SV40 T antigen-binding sites. *Cell* 8:535–545, 1976.

41.   Davis R, Simon M, Davidson N: Electron microscope heteroduplex methods for mapping regions of base sequence homology in nucleic acids, in Grossman L, Moldave K (eds): *Methods in Enzymology* QU 29 M56. New York, Academic Press, 1971, vol 21, pp 413–428.

42.   Padgett BL, Walker DL, ZuRhein GM, et al: Cultivation of papova-like virus from human brain with progressive multifocal leucoencephalopathy. *Lancet* 1:1257–1260, 1971.

43.   ZuRhein GM, Chou SM: Particles resembling papova virus in human cerebral demyelinating disease. *Science* 148:1477–1479, 1965.

44.   Weiner LP, Herndon RM, Narayan O, et al: Isolation of virus related to SV40 from patients with progressive multifocal leukoencephalopathy. *N Engl J Med* 286:385–389, 1972.

45.   Sack GH, Narayan O, Danna KJ, et al: The nucleic acid of an SV40-

like virus isolated from a patient with progressive multifocal leukoencephalopathy. *Virology* 51:345–350, 1973.

46. Fareed GC, Byrne JC, Martin MA: Triplication of a unique genetic segment in a simian virus 40-like virus of human origin and evolution of new viral genomes. *J Mol Biol* 86:275–288, 1974.

47. Dougherty RM, Destefano HS: Isolation and characterization of papovavirus from human urine. *Proc Soc Exp Biol Med* 146:481–487, 1973.

48. Narayan O, Penney JB Jr., Johnson RT, et al: Etiology of progressive multifocal leukoencephalopathy. *N Engl J Med* 289:1278–1282, 1973.

49. Gardner SD, Field AM, Coleman DV, et al: New human papovavirus (BK) isolated from urine after renal transplantation. *Lancet* 1:1253–1257, 1971.

50. Field AM, Gardner SD, Goodbody RA, et al: Identity of a newly isolated human polyomavirus from a patient with progressive multifocal leucoencephalopathy. *J Clin Pathol* 27:341–347, 1974.

51. Takemoto KK, Mullarkey MF: Human papovavirus, BK strain: Biological studies including antigenic relationship to simian virus. *J Virol* 12:625–631, 1973.

52. Mullarkey MF, Hruska JF, Takemoto KK: Comparison of 2 human papovaviruses with simian virus 40 by structural protein and antigen analysis. *J Virol* 13:1014–1019, 1974.

53. Lecatsas G, Prozesky OW, Van Wyk J, et al: Papovavirus in urine after renal transplantation. *Nature* 241:343–344, 1973.

54. Jung M, Krech U, Price PC, et al: Evidence of chronic persistent infections with polyomaviruses in renal transplant recipients. *Arch Virol* 47:39–46, 1975.

55. Takemoto KK, Rabson AS, Mullarkey MF, et al: Isolation of papovaviruses from brain tumor and urine of a patient with Wiskott-Aldrich syndrome. *J Natl Cancer Inst* 53:1205–1207, 1974.

56. Shah KV: Evidence for an SV40-related papovavirus infection of man. *Am J Epidemiol* 95:199–206, 1972.

57. Shah KV, Daniel RW, Warszawski R: High prevalence of antibodies to BK virus, an SV40-related papovavirus, in residents of Maryland. *J Infect Dis* 128:784–787, 1973.

58. Mantyjarvi RA, Meurman OH, Vihma L, et al: A human papovavirus (BK), biological properties and seroepidemiology. *Ann Clin Res* 5:283–287, 1973.

59. Gardner SD: Prevalence in England of antibody to human polyomavirus (BK). *Br Med J* 1:77–78, 1973.

60. Padgett BL, Walker DL: Prevalence of antibodies in human sera against JC virus, an isolate from a case of progressive multifocal leukoencephalopathy. *J Infect Dis* 127:467–470, 1973.

61. Black PH, Rowe WP, Turner HC, et al: A specific complement-fixing antigen present in SV40 tumor and transformed cells. *Proc Natl Acad Sci* 50:1148–1156, 1963.

62. Tegtmeyer P: Simian virus 40 deoxyribonucleic acid synthesis: The viral replicon. *J Virol* 10:591–598, 1972.

63. Tegtmeyer P, Ozer HL: Temperature-sensitive mutants of simian virus 40: Infection of permissive cells. *J Virol* 8:516–524, 1971.

64.   Chou JY, Avila J, Martin RG: Viral DNA synthesis in cells infected by temperature-sensitive mutants of simian virus 40. *J Virol* 14:116–124, 1974.

65.   Chou JY, Martin RG: Infectivity and the induction of host DNA synthesis with temperature-sensitive mutants of simian virus 40. *J Virol* 15: 145–150, 1975.

66.   Martin RG, Chou JY: Simian virus 40 functions required for the establishment and maintenance of malignant transformation. *J Virol* 15: 599–612, 1975.

67.   Tegtmeyer P: Function of simian virus 40 gene A in transforming infection. *J Virol* 15:613–618, 1975.

68.   Brugge HS, Butel JS: Role of simian virus 40 gene A function in maintenance of transformation. *J Virol* 15:619–635, 1975.

69.   Osborn M, Weber K: Simian virus 40 gene A function and maintenance of transformation. *J Viol* 15:636–644, 1975.

70.   Martin RG, Persico-Dibauro M, Edwards, CAF, et al: The molecular basis of transformation by SV40: The semiautonomous replicon revisited. *Proc Int Union Biochem,* Varnasi, India, 1976.

71.   Girardi AJ, Defendi V: Induction of SV40 transplantation antigen (TrAg) during the lytic cycle. *Virology* 42:688–698, 1970.

72.   Anderson JL, Martin RG, Chang C, et al: Tumor specific transplantation antigen is expressed during SV40 lytic infection with wild-type and TsA mutant virus. *Virology* 76:254–262, 1977.

73.   Tenen DG, Martin RG, Anderson JL, et al: Biological and biochemical studies of cells transformed by simian virus 40 temperature sensitive gene A mutants and A mutant revertants. *J Virol* 22:210–218, 1977.

74.   Padgett BL, Hunt JM, Walker DL: Specificity of the tumor-specific transplantation antigen induced by JC virus, a human polyomavirus. *Intervirology* 8:182–185, 1977.

75.   Albert A, ZuRhein GM: Application of immune electron microscopy to the studies of the antigen relationships between 3 new human papovaviruses. *Int Arch Allergy Appl Immunol* 46:405–416, 1974.

76.   Mason DA, Takemoto KK: Complementation between BK human papovavirus and a simian virus 40 tsA mutant. *J Virol* 17:1060–1062, 1976.

77.   Howley PM, Mullarkey MF, Takemoto KK, et al: Characterization of human papovavirus BK DNA. *J Virol* 15:173–181, 1975.

78.   Osborn JE, Robertson SM, Padgett BL, et al: Comparison of JC and BK human papovaviruses with simian virus 40: Restriction endonuclease digestion and gel electrophoresis of resultant fragments. *J Virol* 13:614–622, 1974.

79.   Howley PM, Khoury G, Byrne JC, et al: Physical map of the BK virus genome. *J Virol* 16:959–973, 1975.

80.   Khoury G, Howley PM, Garon C, et al: Homology and relationship between the genomes of papovaviruses, BK virus and simian virus 40. *Proc Natl Acad Sci* 72:2563–2567, 1975.

81.   Shah KV, Ozer HL, Ghazey HN, et al: Common structural antigen of papovaviruses of the SV40-polyoma subgroup. *J Virol* 21:179–186, 1977.

82.   Howley PM, Khoury G, Takemoto KK, et al: Polynucleotide sequences common to genomes of simian virus 40 and the human papovaviruses JC and BK. *Virology* 73:303–307, 1976.

83.   Osborn JE, Robertson SM, Padgett BL, et al: Comparison of JC and BK human papovaviruses with simian virus 40: DNA homology studies. *J Virol* 19:675–684, 1976.

84.   Shah KV, Daniel RW, Strandberg JD: Sarcoma in a hamster inoculated with BK virus, a human papovavirus. *J Natl Cancer Inst* 54:945–950, 1975.

85.   Walker DL, Padgett BL, ZuRhein GM, et al: Human papovavirus (JC): Induction of brain tumors in hamsters. *Science* 181:674–676, 1973.

86.   Major EO, diMayorca G: Malignant transformation of BKH_{21} clone 13 cells by BK virus—a human papovavirus. *Proc Natl Acad Sci* 70:3210–3212, 1973.

87.   Portolani M, Bardanti-Brodano G, LaPlaca M: Malignant transformation of hamster kidney cells by BK virus. *J Virol* 15:420–422, 1975.

88.   van der Noordaa J: Infectivity, oncogenicity, and transforming ability of BK virus and BK virus DNA. *J Gen Virol* 30:371–373, 1976.

89.   Costa J, Howley PM, Legallais F, et al: Oncogenicity of a nude mouse cell line transformed by a human papovavirus. *J Natl Cancer Inst* 56:863–864, 1976.

90.   Takemoto KK, Martin MA: Transformation of hamster kidney cells by BK papovavirus DNA. *J Virol* 17:247–253, 1976.

91.   Weiss AF, Portmann R, Fischer H, et al: Simian virus 40-related antigens in three human meningiomas with defined chromosome loss. *Proc Natl Acad Sci* 72:609–613, 1975.

92.   Soriano F, Shelburne CE, Göcken M: Simian virus 40 in a human cancer. *Nature* 249:421–424, 1974.

93.   Fiori M, diMayorca G: Occurrence of BK virus DNA in DNA obtained from certain human tumors. *Proc Natl Acad Sci* 73:4662–4666, 1976.

94.   Holmberg CA, Gribble DH, Takemoto KK, et al: Isolation of SV40 from spontaneous multifocal leukoencephalopathy in rhesus monkeys. *Infect Immun,* in press.

95.   Khoury G, Carter BJ, Ferdinand FJ, et al: Genome localization of simian virus 40 RNA species. *J Virol* 17:832–840, 1976.

12

RNA Tumor Viruses

PAUL H. BLACK

A large body of accumulated evidence indicates that viruses are the causative agents of some animal cancers. The viruses that have been most intensively studied are the oncogenic RNA viruses, that is, the oncornaviruses or retroviruses. These viruses are responsible for some animal leukemias, at least in mice, chickens, and cats, and for mammary carcinoma in certain inbred strains of mice. Studies have also shown that certain viruses can transform cells *in vitro,* a single virus particle being sufficient to cause one transformation event. Furthermore, utilizing temperature-sensitive mutants of RNA sarcoma viruses and *in vitro* transformation systems, it has been shown that a single virus gene product is sufficient to initiate neoplastic transformation and to maintain the transformed phenotype.

These findings have stimulated an intensive search for a human cancer virus. Within the past 15 years, many claims have been made for the association of some virus or viral product with human malignancy. Such claims have been based on several criteria, which include (a) data obtained by electron microscopy; (b) the finding of RNA-dependent DNA polymerase (reverse transcriptase) activity in certain cancers; (c) the presence of particles that have the same biophysical properties as viruses in or elaborated by the cancer cells; and (d) the presence of viral antigens in the cancer cells.

Most of these reports are controversial, and authorities do not yet agree about any RNA virus as the possible cause of human cancer. The human cancers that have been studied most extensively with regard to possible oncornavirus cause are leukemias, breast cancer, and sarcomas. I will focus on the first two of these by examining the evidence that viruses cause these cancers. In this connection, two specific questions may be asked:

1. Can endogenous viruses cause neoplastic transformation in the host of origin?
2. Can endogenous viral genes become derepressed as the cancer progresses?

HUMAN BREAST CANCER

Many analogies have been made between human breast cancer and the tumors caused by the mouse mammary tumor virus (MMTV), which I shall therefore briefly review. MMTV is an RNA virus of the B type and has characteristic surface projections, or spikes (1). The agent exists as a provirus in the cellular DNA of nearly all strains of mice. The virus is transmitted to neonatal mice by the mother's milk. In strains of mice with high incidence of mammary tumors, large amounts of virus are acquired, and tumors develop early. If mice of the high incidence strains are foster-nursed on virus-negative dams, the appearance of milk-borne virus as well as tumors is delayed. Virus contained in the tumors is termed the "genome virus" and is apparently less virulent and has a lower tumor-inducing potential than the virus in the milk (2,3). Besides the virus that may be necessary for tumor development, genetic factors determine susceptibility or resistance to cancer of a particular strain, and progesterone, prolactin, and estrogenic hormones are apparently essential for tumor development (4). Generally, large amounts (approximately 10^{10} to 10^{12} particles) of virus are present in the milk of high-incidence strains of mice, in which virus can readily be seen in sections of tumors and explanted tumors may release infectious virus into the culture fluid. Interestingly, in strains of mice that are "virus-negative" and have a low incidence of mammary cancer, such as the C57B1 strain, large amounts of viral RNA may be present in mammary tissue, especially in lactating glands (5). The amount of RNA per cell may actually be greater than in Balb C mice, in which the incidence of mammary cancer is appreciable late in life. Thus, virus expression per se, even in hormonally stimulated glands, is not sufficient to cause mammary neoplasia. MMTV sequences have not been found in the DNA of other, closely related rodent species such as the rat, hamster, or European field vole (5). Viruses are "associated" with breast cancer in two other mammalian species, namely, the rat and the rhesus monkey; but the Mason-Pfizer virus, isolated from the latter species, has no nucleic acid homology with the MMTV (1).

With the MMTV as a model, the search for a virus in human breast cancer has been intense. Several types of "particles" have been seen in human milk, only one of which resembles the B particle. The other types of particle have surface spikes unlike those of the B particle or resemble the Mason-Pfizer monkey virus (6). Particles with B-like morphology are exceedingly rare, with perhaps one or two in each milk specimen being seen after prolonged search (1). In general, little success has been achieved in obtaining infectious material from either human milk or human mammary cancer. In one study, however, human embryonic cell lines, on cocultivation with human breast cancer cells or inoculated with milk from breast cancer patients, developed cytopathology and were destroyed. Moreover, particles were recovered that had a density of 1.16–1.19 gm/ml, contained 70S RNA as well as RNA-dependent DNA polymerase activity (7), and had the biophysical properties of oncogenic viruses.

In other studies, the presence of particles in human milk has been reported, or particles that have the density of virions (1.16–1.19 gm/ml) have been obtained from breast cancer tissue extracts. Reverse transcriptase activity is associated with the particles that contain RNA of high molecular weight (8–10). Radiolabeled complementary DNA (cDNA) probes made from the RNA in such particles may react with polysomal RNA from human breast cancers (8). Of additional interest is the observation that cDNA probes made from MMTV were found to hybridize with the RNA of human breast tumors (18% to 77% of a MMTV cDNA probe at C_rt values of 2.2 to 5.1 $\times$ 10^4 mole s/liter) (11), and 67% of 29 human breast tumor polysomal RNA preparations hybridized to a significant extent with a similar probe but not with a cDNA probe prepared from Rauscher leukemia virus (RLV) (12). It has also been claimed that a cDNA probe prepared from the human breast "particle" will hybridize to the RNA of MMTV (12).

MMTV antigens have also been detected in human materials. By indirect immunofluorescence tests, sera from patients with breast cancer and from relatives of breast cancer patients were found to react with mouse tissue culture cells, which were producing B- or C-type particles. However, 15% of all control sera gave positive reactions, a good part of which was ascribed to Forssmann heterophile antigens (13). Other studies indicated that human sera from breast cancer patients neutralized MMTV, whereas normal sera did not (14). Later, however, normal human sera were also found to neutralize MMTV (15). There is, however, some evidence that MMTV-related antigens may be present in human material; antigens prepared from a large pool of human milk samples were shown by micro-Ouchterlony and radioimmunoassays to be similar to the MMTV gp 55 glycoprotein (15).

An agent has been recovered from a cell line that had originated from a pleural effusion from a patient with disseminated adenocarcinoma of the breast; the cells retained properties characteristic of mammary epithelium, such as estrogen receptors and α-lactalbumin synthesis (16). The particle isolated from the supernatant fluid of these cultures had a density of 1.16 to 1.18 gm/ml and contained RNA with a molecular weight of 10 to 12 $\times$ 10^6 daltons as well as reverse transcriptase. Antiserum made against detergent-solubilized MMTV structural proteins, glycoprotein (gp) 52, protein (p) 28, p 15, p 10 to 12, revealed antigens present in the cell that were indistinguishable by immunofluorescence from MMTV antigens. Hybridization occurred with a cDNA probe made against several MMTV RNAs and RNA from the cell line. Reactions also occurred with a cDNA made from particles elaborated by this cell line and normal human DNA under stringent hybridization conditions. The data suggest that genes in the viral particles may have originated from human DNA.

The evidence presented suggests that "viruses" similar in biophysical properties and antigenicity to, and related genetically to, mouse B- or C-type viruses are associated with human mammary neoplasia and may be present in normal breast milk. Moreover, the results of studies by McGrath et al. (16) suggest that the agent is an endogenous human virus. However, MMTV-induced mammary cancer in mice differs considerably from carcinoma of the breast in humans. Except for consistent familial aggregation, there is only meager epidemiologic evidence for a viral cause of breast cancer.

Observations regarding the possible viral causation of human breast cancer may be summarized as follows: The particles seen by electron microscopy are heterogeneous, B-type, C-type, and MP simian virus having been described in the same impure preparation (6), and the number of milk particles found in normal subjects is often the same as that found in cancer patients (6). In most studies, no biologic activity has been demonstrably associated with milk particles. Reports vary as to whether MP simian virus (17) or its antigens (18) are present in human milk. According to one report, no correlation was found between milk fractions having reverse transcriptase activity and those containing MP-like particles (17). With one exception (19), no particles have been observed either in breast cancer tissue or in cultures derived from such tissue (20). Apparent antigenic cross-reactions between MMTV and breast cancer tissue were due to nonspecific factors (13). Moreover, virus neutralization could be demonstrated with sera from cancer patients as well as with normal sera, perhaps as a result of the lytic effect of complement, as recently demonstrated with C-type viruses (21).

Human breast cancer tissue may contain DNA homologous to

MMTV; however, the significance of this is not clear. Some data tend to indicate that the MMTV-related agent may be an endogenous human virus, but nucleic acid sequences homologous to MMTV have not been found, even in closely related rodent species such as the hamster and rat (5). Moreover, nucleic acid hybridization studies failed to show the presence of MMTV-related sequences in normal human DNA (11,12). If the putative human mammary cancer agent is not endogenous, then it must be acquired horizontally, either through milk or by contagion. Little evidence now exists to support either mode of acquisition of the putative virus. In murine oncogenesis, MMTV acts in a genetically manipulated inbred population under unique circumstances (1), and the virus of greatest oncogenicity is one that is ingested in large quantities during suckling, a factor that itself is presumably important in the pathogenesis of murine mammary cancer. Thus far, no comparable situation has been shown to exist in humans.

HUMAN LEUKEMIA

The search for a virus in human malignancy has been most intensive with human leukemias and lymphomas because there is strong evidence for a causal relationship between viruses and leukemia in various animal species, that is, chickens, cats, cows, outbred mice, and gibbon apes. In those animals in which virus is usually acquired horizontally, virus replicates to high titer and is therefore easily isolated. One major question is whether the endogenous viruses whose genetic information is resident in the host genome, if activated, can give rise to a neoplastic transformation event.

Enzymes that catalyze the synthesis of DNA on an RNA template (reverse transcriptase) similar to those in oncornaviruses have been described in human leukemic tissue. The enzyme activity was found to be associated with human acute myelocytic leukemic cells and to be immunologically related to reverse transcriptase of simian sarcoma virus (SSV) isolated from a lymphosarcoma of the Woolly monkey (22). A similar enzyme was isolated from the cytoplasm of peripheral leukocytes of a patient with myelomonocytic leukemia. This human leukocyte enzyme, after extensive purification, was inhibited by antisera prepared against the reverse transcriptase of SSV (23) and was found to utilize synthetic primer-templates as did the simian viral enzyme (24). This type of enzymatic activity, however, has been found only rarely in leukocytes of acute myelocytic leukemic patients. Many other enzymes from human leukemic tissue have been described as having viruslike properties but without any of the biophysical or immunologic properties of viral reverse transcriptase (25). In fact, some react like normal cellular DNA polymerases (26).

Several studies have been done with "particles" isolated from fresh tumor tissue that band at densities characteristic of C-type viruses and contain high molecular weight RNA (60S to 70S) as well as reverse transcriptase activity. These particles, however, do not replicate and show no biologic activity. Radiolabeled cDNA synthesized from the particles *in vitro* by endogenous reverse transcriptase can be hybridized to the RNA of some mammalian viruses. By such techniques cDNA derived from human leukemic particles is shown to bear homology with the RNA of RLV but not with the RNA of either avian myeloblastosis virus or MMTV (27,28). A cDNA synthesized from isopycnically purified cytoplasmic particles was shown to hybridize to SSV virus (26,29–31) and to Kirsten sarcoma virus (26,29). In a few instances, particles purified from short-term cultures were seen by electron microscopy to resemble oncornaviruses (31,32). Reverse transcriptase activity associated with the particles increased after culture, and two discrete bands of particles were visualized at densities, respectively, of 1.17 and 1.23 gm/ml. Particles were obtained from patients with acute or chronic myelocytic leukemia and acute lymphocytic leukemia, in both remission and relapse. Another report describes the liberation of C-type particles after cocultivation of X-C cells with human adenocarcinoma cells from a patient with chronic lymphatic leukemia. By nucleic acid homology and immunologic studies, the virus was found to be similar to SSV (33).

Another approach has been to hybridize a cDNA copy of the genome of RNA viruses with tumor cell RNA to determine whether genetic information related to that of known viruses is present in tumor cells. The cDNA probes synthesized to match a murine sarcoma virus hybridize more readily to the RNA of leukemia tissue than to RNA of normal tissue (34). Similar results have been found with cDNA prepared from SSV (26). A cDNA probe from RLV has been found to hybridize with either a polysomal or cytoplasmic RNA fraction of human sarcomas (35), human leukemia (36), Hodgkin's disease and other lymphomas (37), Burkitt's lymphoma, and nasopharyngeal carcinoma (38). Whether the same Rauscher RNA sequences are responsible for the purported homology in all these cancers is not known. Usually the amount of cDNA hybridized in such experiments is small, and the cell RNA is so complex that the results may be difficult to interpret in terms of disclosing the presence of a particular viral RNA. Thus far, the relevant RNA has not been purified from any human tumor tissue.

In the studies mentioned above, the particles isolated from tumor tissue or released after short-term tissue culture were not infectious, and budding from the cell was not observed. Infectious virus related to the simian sarcoma agent was isolated after 4 to 10 passages from cultured blood cells from a patient with acute myelogenous leukemia (39). The virus could be propagated *in vitro* (40). It was reisolated from a bone marrow specimen obtained from the same patient 14 months later (41). Subsequently, it was shown that the virus obtained was a mixture of SSV and the baboon endogenous virus (30,42,43). Interestingly, sequences homologous to the baboon endogenous virus are present in the spleens of leukemic patients to a greater extent than in the spleens of nonleukemic persons (44). However, whether the additional amount of endogenous baboon proviral DNA represents amplification or the acquisition of new sequences is not certain. Also of interest is the fact that the virus isolated from normal human embryo cells (45) is also a mixture of SSV and the endogenous baboon virus (46). Nucleic acid hybridization data indicate that normal human DNA contains sequences homologous to a portion of the baboon endogenous virus (47); no sequences homologous to SSV have been detected in normal human DNA, however (48).

CRITIQUE

Nucleic acid sequences in human cancers may be found that relate to mouse and primate viruses and are associated with cytoplasmic particles, with elaborated particles, or, more rarely, with budding particles. With few exceptions, the nature of the particles remains obscure. It is clear that little or no homology exists between normal human DNA and any mouse C-type viral RNA, and there is no serologic evidence of human infection by any known mouse oncogenic viruses. With regard to the primate isolates, the situation is more complicated. As stated, normal human DNA contains sequences homologous to the baboon endogenous virus (47). However, no homology between normal DNA and SSV has been found. Therefore, SSV probably does not represent an endogenous human virus (48). There is some uncertainty as to whether humans can become infected with this agent. Results of immunoelectron-microscopic studies have indicated that normal human sera contain antibodies to SSV and the endogenous baboon viruses (49) but otherwise failed to detect any antibody to SSV by radioimmunoassay (50) with purified viral structural proteins. Moreover, in the latter studies, SSV could not be induced from any of more than 200 cultures of normal human cells by treatment with halogenated pyrimidines, further indicating that SSV is unlikely to be an endogenous human virus (50).

One must be skeptical about the data that have accumulated thus far relating viruses to either breast cancer or leukemia in humans. It has not been proved conclusively that the putative human cancer proviral sequences are confined to the malignant tissue of cancer patients or even to patients with cancer (25). The rarity of the findings presented is noteworthy. Accordingly, I do not mean to imply that all cancers of a certain type must be caused by viruses. Certainly, leukemia may have several causes; radiation and chemicals are known to cause leukemia, and it is conceivable that viruses may also do so. However, it is unlikely that viruses are the direct cause in all cases and that the stimuli mentioned act by activating viruses as proposed by the oncogene theory (51). If the mouse and SSV virus sequences present in human cancer are real, and if, as the evidence indicates, these agents are not endogenous in humans, then they can only be acquired horizontally. However, there is no epidemiologic evidence to indicate that a virus causes human leukemias, although a prenatal origin has been suggested by some. In the latter instance the data are also compatible with pathogenetic mechanisms other than viral. Evidence for *in utero* transmission of a pathogenic agent is scanty, and children born to leukemic mothers are not at increased risk of the disease. No seasonal trend has been noted in the onset of human leukemia or in the births of children in whom leukemia develops. Leukemia risks do not seem to be increased in spouses of leukemic patients or among owners of leukemic dogs or cats. Some reports of time-space clustering of the disease have appeared, but the number of negative studies far exceeds that of the positive, among which the clustering is only slight (52).

One cannot readily explain the few positive findings I have cited. The finding of a virus-related reverse transcriptase is exceedingly rare when one applies stringent criteria (25), and "particles" present in tumor cells have generally not been rigorously purified or preparatively isolated for examination. The cytoplasm of a tumor cell can synthesize DNA on a high molecular weight RNA template, whereas normal tissues cannot. Although this cytoplasmic capability is a measure of cDNA synthesis, it may reflect reverse transcriptase activity associated with endogenous viral genetic expression. The assay, however, is indirect, and the reaction product has not been analyzed in detail (25). The methods for analysis of hybrid formation have not always been critical, and the counts of the hybridized material may sometimes be low.

The presence in human DNA of sequences similar to baboon DNA indicates that humans have endogenous virus genetic information, and the finding of additional proviral sequences or viral RNA in a tumor may represent gene amplification or spontaneous induction of an endogenous viral genome. The induction conceivably could result from the development of a cancer and represent gradual derepression of genetic information accompanying progression of the tumor. The expression of fetal genetic information, that is, the "fetalization" of a tumor, is an example of such a process. Could the expression of endogenous viral genetic information result in malignant transformation? This seems unlikely because endogenous viruses, at least those of chickens, cats, and baboons, have not been shown to be oncogenic, and from the presently available evidence the same is probably true of the human endogenous virus.

The statement is often made that a putative human cancer virus may be defective and that the cancer cell may harbor the viral genome in a nonproducer state. Such a situation would be analogous to that underlying avian, feline, and murine sarcoma viruses, which can be "rescued" by certain leukemia viruses. This type of rescue with human cancer cells has been attempted many times but with little success. Possibly a complete viral genome exists in a cancer cell without production of infectious virus, that is, the virogenic state (53). Such viruses may be rescued by various inductive stimuli, such as mitomycin C, radiation, or halogenated pyrimidines. Again, such measures have met with little success with human cancer tissue. It seems more likely that a more conventional type of virus infection would be involved if human cancer were caused by a virus. Indeed, the animal cancers caused by viruses are likely to involve large amounts of horizontally transmitted viruses: the Bittner MMTV is present in large amounts in milk; cat, bovine, and chicken leukemias also involve large amounts of horizontally transmitted viruses. When an endogenous virus is the cause of an animal cancer, large amounts of the virus are usually present, such as the Gross virus in AKR mouse leukemia. An analogous situation has not been found to occur in humans. One should also keep in mind that the experimental systems that have been most studied involve inbred genetically uniform strains of animals. Accordingly, mechanisms operative with the MMTV and the Gross AKR viruses in mice may not be applicable to humans, a highly outbred species. Nevertheless, the quest for a human cancer virus will and should continue.

DISCUSSION

Question: What is the relationship of the human C-type virus of Panem and Kirsten to systemic lupus erythematosus?

Dr. Black: These investigators have described antigens related to SSV and the endogenous baboon virus in the immune complexes in kidneys of patients with disseminated lupus erythematosus (DLE). These viruses are the same two viruses that the same investigators have "activated" from human diploid fibroblast cultures and that R. C. Gallo and colleagues have isolated from cells from a patient with acute myelogenous leukemia.

Whether these viruses are causally related to the DLE is unknown. The viral specificity of the immunofluorescent staining of the immune complexes in DLE has not been unambiguously determined. It is likely that humans have an endogenous type C virus, and it might become activated in lupus, perhaps from the B lymphocytes that are apparently released from T cell suppressor restraint. It is known that activated lymphocytes may release endogenous viruses (see ref. 52). This situation would be analogous to the release of endogenous viruses from cancer cells, in which genetic repressive mechanisms are generally abrogated. The presence of SSV cannot be explained on this basis, however, and its presence in DLE, if real, is as unexplained as its presence in human leukemia, if real.

REFERENCES

1. Moore DH: Mammary tumor virus, in Becker FF (ed): *Cancer, A Comprehensive Treatise.* New York, Plenum Press, 1975, vol 2, pp 131–168.
2. Moore DH: Mouse mammary tumor agent and mouse mammary tumors. *Nature* 198:429–433, 1963.
3. Bentvelzen P, Daams JH, Hageman P, et al: Genetic transmission of viruses that incite mammary tumor in mice. *Proc Natl Acad Sci* 67:377–384, 1970.
4. Muhlbock O: Role of hormones in etiology of breast cancer. *J Natl Cancer Inst* 48:1213–1216, 1972.
5. Varmus HE, Quintrell N, Medeiros E, et al: Transcription of mouse mammary tumor virus genes in tissues from high and low tumor incidence mouse strains. *J Mol Biol* 79:633–679, 1973.
6. Sarkar NH, Moore DH: On the possibility of a human breast cancer virus. *Nature* 236:103–106, 1972.
7. Keydar J, Gilead Z, Karby S, et al: Production of virus by embryonic cultures cocultivated with breast tumor cells or infected with milk from breast cancer patients. *Nature (New Biol)* 241:49–52, 1973.
8. Das MR, Sadisivan E, Koshy R, et al: Homology between RNA from human malignant breast tissue and DNA synthesized by milk particles. *Nature* 239:92–95, 1972.
9. Schlom J, Spiegelman S, Moore DH: RNA dependent DNA polymerase activity in virus-like particles isolated from human milk. *Nature* 231:97–100, 1971.
10. Schlom J, Spiegelman S, Moore DH: Reverse transcriptase and high molecular weight RNA in particles from mouse and human milk. *J Natl Cancer Inst* 48:1197–1203, 1972.
11. Vaidya AB, Black MM, Dion AS, et al: Homology between human breast tumor RNA and mouse mammary tumor virus genome. *Nature* 249:565–567, 1974.
12. Axel R, Gulati C, Spiegelman S: Particles containing RNA instructed DNA polymerase and virus-related RNA in human breast cancer. *Proc Natl Acad Sci* 69:3133–3137, 1972.
13. Priori ES, Anderson DE, Williams WC, et al: Immunological studies on human breast carcinoma and mouse mammary tumors. *J Natl Cancer Inst* 48:1131–1135, 1972.
14. Charney J, Moore DH: Neutralization of murine mammary tumor viruses by sera of women with breast cancer. *Nature* 229:627–628, 1971.
15. Moore DH: Viruses and breast cancer, in Crowell RL, Friedman H, Prier JE (eds): *Tumor Virus Infections and Immunity.* Baltimore, University Park Press, 1976, pp 45–61.
16. McGrath CM, Furmanski P, Russo J, et al: 734B: A candidate human breast cancer virus, in Crowell RL, Friedman H, Prier JE (eds): *Tumor Virus Infections and Immunity.* Baltimore, University Park Press, 1976, pp 63–87.
17. Chopra H, Ebert P, Woodside N, et al: Electron microscopic detec-

tion of simian-type virus particles in human milk. *Nature* 243:159–160, 1973.

18. Yeh J, Ahmed M, Mayyasi SA, et al: Detection of an antigen related to Mason-Pfizer virus in malignant human breast tumors. *Science* 190:583–584, 1975.

19. Seman G, Myers B, Williams WC, et al: Studies on the relationship of viruses to the origin of human breast cancer: II. Virus-like particles in human breast tumors. *Texas Rep Biol Med* 27:839–866, 1969.

20. Cailleau R, Young R, Olive M, et al: Breast tumor cell lines from pleural effusions. *J Natl Cancer Inst* 53:661–674, 1974.

21. Jensen FC, Welsh RM, Cooper NR, et al: Lysis of oncornaviruses by human serum, in Clemmensen J, Yohn DS (eds): *Bibl Haematol* No 43. Basel, Karger, 1976, pp 438–440.

22. Bhattacharyya J, Xuma M, Reitz M, et al: Utilization of mammalian 70S RNA by a purified reverse transcriptase from human myelocytic leukemic cells. *Biochem Biophys Res Comm* 54:324–334, 1973.

23. Gallagher RE, Todaro GJ, Smith RG, et al: Relationship between RNA-directed DNA polymerase (reverse transcriptase) from human acute leukemic blood cells and primate type C viruses. *Proc Natl Acad Sci* 71:1309–1313, 1974.

24. Mondal H, Gallagher RE, Gallo RC: RNA-directed DNA polymerase from human leukemic blood cells and from primate type-C virus-producing cells: High and low molecular weight forms with variant biochemical and immunological properties. *Proc Natl Acad Sci* 72:1194–1198, 1975.

25. Gallo RC, Gillespie DH: The origin of RNA tumor viruses and their relation to human leukemia, in Silber RD, Gordon AS, Lobue J (eds): *The Yearbook of Hematology* 1976–1977. New York, Plenum Press, 1977.

26. Gallo RC, Gallagher RE, Miller NR, et al: Relationships between components in RNA tumor viruses and in the cytoplasm of human leukemic cells: Implications to leukemogenesis. *Cold Spring Harbor Symp Quant Biol* 34:933–961, 1974.

27. Baxt W, Hehlmann R, Spiegelman S: Human leukemic cells contain reverse transcriptase associated with a high molecular weight virus-related RNA. *Nature (New Biol)* 240:72–75, 1972.

28. Baxt W, Speigelman S: Nuclear DNA sequences present in human leukemic cells and absent in normal leukocytes. *Proc Natl Acad Sci* 69:3737–3741, 1972.

29. Gallo RC, Miller NR, Saxinger WC, et al: Primate RNA tumor virus-like DNA synthesized endogenously by RNA-dependent DNA polymerase in virus-like particles from fresh human acute leukemic blood cells. *Proc Natl Acad Sci* 70:3219–3224, 1973.

30. Reitz MS, Miller NR, Wong-Staal F, et al: Primate type C virus nucleic acid sequences (Woolly monkey and baboon types) in tissues from a patient with acute myelogenous leukemia and in viruses isolated from cultured cells of the same patient. *Proc Natl Acad Sci* 73:2113–2117, 1976.

31. Mak TW, Kurtz S, Manaster J, et al: Viral-related information in oncornavirus-like particles isolated from cultures of marrow cells from leukemic patients in relapse and remission. *Proc Natl Acad Sci* 72:623–627, 1975.

32. Mak TW, Manaster J, Howatson AF, et al: Particles with characteris-

tics of leukoviruses in cultures of marrow cells from leukemic patients in relapse and remission. *Proc Natl Acad Sci* 71:4336–4340, 1974.

33. Gebelman N, Waxman S, Smith W, et al: Appearance of C-type virus-like particles after cocultivation of a human tumor-cell line with rat (X–C) cells. *Int J Cancer* 16:355–369, 1975.

34. Larsen CJ, Marty M, Hamelin R, et al: Search for nucleic acid sequences complementary to a murine oncornaviral genome in poly (A)-rich RNA of human leukemic cells. *Proc Natl Acad Sci* 72:4900–4904, 1975.

35. Kufe D, Hehlmann R, Spiegelman S: Human sarcomas contain RNA related to the RNA of a mouse leukemia virus. *Science* 175:182–185, 1972.

36. Hehlmann R, Kufe D, Spiegelman S: RNA in human leukemic cells related to the RNA of a mouse leukemia virus. *Proc Natl Acad Sci* 69:435–439, 1972.

37. Hehlmann R, Kufe D, Spiegelman S: Viral-related RNA in Hodgkin's disease and other human lymphomas. *Proc Natl Acad Sci* 69:1727–1731, 1972.

38. Kufe D, Hehlmann R, Spiegelman S: RNA related to that of a murine leukemia virus in Burkitt's tumors and nasopharyngeal carcinomas. *Proc Natl Acad Sci* 70:5–9, 1973.

39. Gallagher RE, Gallo RC: Type C RNA tumor virus isolated from cultured human acute myelogenous leukemia cells. *Science* 187:350–353, 1974.

40. Teich NM, Weiss RA, Salahuddin SZ, et al: Infective transmission and characterization of a C-type virus released by cultured human myeloid leukemia cells. *Nature* 256:551–555, 1975.

41. Gallagher RE, Salahuddin SZ, Hall WT, et al: Growth and differentiation in culture of leukemic leukocytes from a patient with acute myelogenous leukemia and reidentification of a type C virus. *Proc Natl Acad Sci* 72:4137–4141, 1975.

42. Okabe H, Gilden RV, Hatanaka M, et al: Immunological and biochemical characterization of type-C viruses isolated from cultured human AML cells. *Nature* 260:264–266, 1976.

43. Chan E, Peters WP, Sweet RW, et al: Characterization of a virus (HL 23V) isolated from cultured acute myelogenous leukemic cells. *Nature* 260:266–268, 1976.

44. Wong-Staal F, Gillespie D, Gallo RC: Proviral sequences of baboon endogenous type C RNA virus in DNA of leukemic tissues from 7 patients with myelogenous leukemia. *Nature* 262:190–195, 1976.

45. Panem S, Prochownik EV, Reale FR, et al: Isolation of C type virions from a normal human fibroblast strain. *Science* 189:297–299, 1975.

46. Prochownik EV, Kirsten W: Inhibition of reverse transcriptase of primate type C viruses by 7S immunoglobulin from patients with leukemia. *Nature* 260:64–67, 1976.

47. Benveniste RE, Todaro GJ: Evolution of type C viral genes: I. Nucleic acid from baboon type C virus as a measure of divergence among primate species. *Proc Natl Acad Sci* 71:4315–4318, 1974.

48. Scolnick EM, Parks W, Kawakami T, et al: Primate and murine type-C viral nucleic acid association kinetics: Analysis of model systems and natural tissues. *J Virol* 13:363–369, 1974.

49. Aoki T, Walling MJ, Bushar GS, et al: Natural antibodies in sera

from healthy humans to antigens on surfaces on type C RNA viruses and cells from primates. *Proc Natl Acad Sci* 73:2491–2495, 1976.

50. Stephenson JR, Aaronson SA: Search for antigens and antibodies crossreactive with type C viruses of the Woolly monkey and gibbon ape in animal models and in humans. *Proc Natl Acad Sci* 73:1725–1729, 1976.

51. Allen DW, Cole P: Viruses and human cancer. *N Engl J Med* 286: 70–82, 1972.

52. Hirsch MS, Black PH: Activation of mammalian leukemia viruses. *Adv Virus Res* 19:265–313, 1974.

53. Kaplan JC, Black PH, Rothschild H: Induction of virogenic mammalian cells with chemical and physical agents. *Can Cancer Conf Proc* 9: 143–158, 1971.

13

Unconventional Viruses

D. CARLETON GAJDUSEK

Do slow viruses really exist? Certainly, viruses are implicated in a group of fatal subacute and chronic diseases that affect both man and animals. Unfortunately, ambiguity results from the American propensity to use nouns as modifiers, whereby many persons do not recognize that, in the term "slow virus infection," "slow" modifies the noun "infection" and not the noun "virus"; for that reason, I prefer to describe such agents as unconventional or atypical and the diseases as slow.

Until about a decade ago, chronic diseases not otherwise classifiable had been labeled simply as degenerative. Virus-caused diseases, on the other hand, were usually considered to be acute and self-limiting, leading quickly either to recovery or to death. However, during the past two decades, knowledge about the infectious cause of certain chronic progressive human diseases of the central nervous system (CNS) has advanced so remarkably that a whole new phase of virology has been opened.

DISEASES CAUSED BY UNCONVENTIONAL VIRUSES

To date, we have demonstrated that four chronic, progressive, and closely related diseases of the CNS, transmissible to animals (Table 13.1), are caused by highly unusual but similar agents (Table 13.2). Two occur in man, kuru and Creutzfeldt-Jakob disease (CJD), and two in animals, scrapie and transmissible mink encephalopathy (TME). We have named these diseases subacute spongiform virus encephalopathies because, in the advanced stage of both natural and

TABLE 13.1. *Naturally Occurring Slow Virus Infections Caused by Unconventional Viruses (Subacute Spongiform Virus Encephalopathies)*

In man:
Kuru
Transmissible virus dementia
Creutzfeldt-Jakob disease
Sporadic
Familial
Familial Alzheimer disease

In animals:
Scrapie
in sheep
in goats
Transmissible mink encephalopathy

TABLE 13.2. *Properties of the Unconventional Viruses*

Physical and Chemical Properties
   Resistant to formaldehyde
        "   " β-propiolactone
        "   " ethylenediamine tetraacetic acid (EDTA)
        "   " proteases (trypsin, pepsin)
        "   " nucleases (RNases A and III, DNase I)
        "   " heat (80°C); incompletely inactivated at 100°C
        "   " ultraviolet radiation: 2540 A
        "   " ionizing radiation (γ rays): equivalent target 150,000 daltons
        "   " ultransonic energy
        "   " desoxycholate
   Atypical ultraviolet action spectrum: 2370 A $= 6 \times$ 2540 A inactivation
   Invisible as recognizable virion by electron microscopy (only plasma membranes, no core and coat)
   No nonhost proteins demonstrated

Biological Properties
   Long incubation period (months to years: decades)
   No inflammatory response
   Chronic progressive pathology (slow infection)
   No remissions or recoveries: always fatal
   "Degenerative" histopathology: amyloid plaques, gliosis
   No visible virion-like structures by electron microscopy
   No inclusion bodies
   No interferon production or interference with interferon production by other viruses
   No interferon sensitivity
   No virus interference (with over 30 conventional different viruses)
   No infectious nucleic acid demonstrable
   No antigenicity
   No alteration in pathogenesis (incubation period, duration, course) by immunosuppression or immunopotentiation:
      (a) ACTH, cortisone
      (b) cyclophosphamide
      (c) x-ray
      (d) antilymphocytic serum
      (e) thymectomy/splenectomy
      (f) "nude" athymic mice
      (g) adjuvants
   Immune "B" cell and "T" cell function intact *in vivo* and *in vitro*
   No cytopathic effect in infected cells *in vitro*
   Varying individual susceptibility to high infecting dose in some host species (as with scrapie in sheep)

TABLE 13.3. *General Criteria of Slow Viral Infections*

---

1. A long initial period of latency, lasting for several months to several years
2. An ultimately fatal illness after the appearance of clinical signs
3. Primary anatomical lesions limited to a single organ system
4. Noninflammatory pathology
5. Absence of immune response
6. Limited range of susceptible hosts
7. Often a heredofamilial pattern of infection

---

experimentally induced disease, status spongiosus, or sponge-like changes appear in the gray matter (1-4).

In 1950, the topic of slow viral infections was not even listed in the *Index Medicus*. A veterinarian, Sigurdsson (5), introduced us to the concept about two decades ago, when he drew attention to four naturally occurring diseases of sheep in Iceland: scrapie, visna, maedi, and pulmonary adenomatosis. These diseases, as well as the human subacute spongiform encephalopathies, fulfill the criteria of slow virus infections (Table 13.3).

Scrapie and Transmissible Mink Encephalopathy

Scrapie, the archetype of slow virus infections, has been recognized in Europe for centuries. The name derives from the behavior of affected sheep, which, apparently because of intense pruritus, denude areas of their skin by rubbing against fences.

Natural scrapie occurs almost exclusively in sheep but has also been found in goats. It may be spread from naturally infected to uninfected sheep, and some breeds of sheep show a high incidence of the disease, whereas others are totally resistant. Because of these differences, genetic control of susceptibility has been postulated (6). The disease seldom affects animals less than two years of age and rarely animals more than five years old. The onset is insidious; early symptoms include slight behavioral changes, such as nervousness and excitability. Later, neurological signs, such as tremor, ataxia, and wasting, following in rapid succession.

In 1899, scrapie was experimentally transmitted by inoculation of a ewe with brain tissue from scrapie-affected sheep. Not until 1936, however, did Cuillé and Chelle (7) confirm that mode of transmission.

No lateral transmission has been observed in experimental sheep or goats. In scrapie, the agent is rarely demonstrable in blood during early or late disease; however, the agent may be found in the spleen, liver, lymph nodes, and other organs as well as in the brain and in the placenta and fetal membranes.

TME, a disease pathologically similar to scrapie, was first described in 1965 (8). TME is somewhat more acute than scrapie, often having a clinical course of only a few days or weeks. The disease is also indistinguishable from that induced in mink by inoculation with tissue from scrapie-affected sheep, and it is caused by an agent that is physicochemically indistinguishable from the scrapie agent.

TME first appeared on mink ranches where carcasses of scrapied sheep were fed to the mink. From the distribution of lesions, as determined clinically and neuropathologically, we were led to believe that scrapie and TME were not related. On the other hand, the basic cytopathological changes seen by electron microscopy are the same in both diseases. It seems probable that this agent, which has all the properties of the scrapie agent, is the scrapie virus modified by passage in mink. When scrapie virus is inoculated experimentally into mink, the agent becomes similarly altered and loses its ability to produce disease in mice, as is the case with natural TME.

Kuru

Clinical Aspects. In 1957 we were studying a strange new subacute fatal brain disease among the Fore people, an isolated mesolithic culture, and their neighbors in the Eastern Highlands of Papua New Guinea. The disease had been present in these endocannibalistic societies for a half century and had reached epidemic proportions. The disease has now virtually disappeared (9). The Fore used their word *kuru* (meaning to shake or shiver) to describe the early symptoms and signs of the disease, namely, a propensity to shiver and shake, which is particularly exaggerated in cold weather.

Restricted to the CNS, the disease has a constant pattern of clinical signs and symptoms. It was common in children, the youngest about age five, but affected adults of all ages. Two-thirds of the patients were adults, mostly women, and one-third were children of both sexes. Often a patient accurately diagnosed the early stage of the disease in himself long before his family was aware of it, and long before any of my colleagues or I were sure that he was affected. Very few who thought they had the disease were wrong in predicting their imminent death from kuru. It invariably began with shivering-like tremor and cerebellar ataxia. Dementia appeared late and was rarely severe. All patients showed ankle clonus and, also, usually spontaneous patellar clonus on standing, and transient hyperreflexia of deep tendon jerks. The lack of muscle control and inability to keep the head erect resulted in a drooping posture; a spastic strabismus appeared, and all patients developed a severe dysarthria. Other indications included pyramidal and extrapyramidal signs, progressing to severe complete motor ataxia and inability to walk or stand or sit unsupported, and death in about one year from onset. Strabismus, found in almost all children, but less frequently in adults, was not sustained, but was spastic and could disappear in 24 hours without ever reappearing, or reappear only to involve a different set of eye muscles. At first we thought we had recorded the observations incorrectly, because in children we found the spastic strabismus often changed sides.

Obesity, a totally unknown phenomenon in the highland population of New Guinea, except in association with kuru, resulted from the victims' overeating and decreased activity because of their ataxia. In our earlier studies, ACTH and cortisone given to some patients contributed to their obese facies. Neither such therapy, nor any other, was effective.

Epidemiology. Kuru was confined to 169 adjacent villages in the mountainous interior of New Guinea. The contiguous areas of fewer than 200 square miles in which the disease occurred involved 11 culturally and linguistically distinct populations (10). We were faced with a problem of exploring the primitive culture by traveling one or two days between villages of fewer than 200 persons. It took us almost two years to reach the entire population of 35,000 to 40,000. Since the beginning of our investigations in 1956, about 3000 cases have been recorded. Our case finding has actually been nearly 100%. By continuous patrolling and living in the villages for the first two years, we covered the whole area. Ever since, we have had surveillance of every human in the area, each of whom was seen on census patrols twice a year.

Kuru was common in children of both sexes and in women, but rare in men. The mechanism of spread was undoubtedly contamination from dead relatives during a cannibalistic rite of respect and mourning. The butchering of dead kinsmen among the Fore was left to the women, and the men and boys rarely ate the flesh of dead kuru victims. In villages where the disease was most common, deaths caused by kuru were frequent enough to ensure that most infants and children were infected during the dissection of the body by the women who were close kinswomen of a deceased relative. The mourning rite, in which contact with the corpse was close and all tissues were cooked and eaten, surely resulted in the infection of most small children. Boys who left the female society at 4 to 6 years of age and did not take part in the cannibalistic rites avoided later contamination. This avoidance of kuru contamination after early childhood by males probably resulted in the observed difference in incidence by sex with disease preponderantly in women.

Kuru was the major cause of death among the Fore. In the South Fore the death rate was barely balanced by the birth rate. In the early days of our study, the deaths of 2% to 3% of the population in some villages annually was common. If such high incidence continued, the village could not survive for long. Villages commonly fissioned and the fragments fused with other, less-affected villages. These changes allowed the remaining population to survive, but also spread the disease by bringing in new members already infected through contamination during the cannibalistic rites of mourning that had occurred before their immigration. Marriages with women already incubating kuru infection further spread the disease. We estimate that the disease was spread to about 10 persons per kuru death.

During the first 15 years of our surveillance in the South Fore area, approximately 90% of all deaths in adult women were caused by kuru. In most of the area the ratio of males to females had reached 2 to 1, and in some villages 3 to 1. For a polygamous people, such male preponderance is a great problem. Were it not for the second commonest cause of death, namely, reprisal murder of suspected male sorcerers, the social situation would have been even less tolerable. Early childhood marriage, with few girls remaining unmarried in the late prepubertal age, was also a good biological adaptation to the disease. Clearly, most girls were pregnant as soon as they could be, and because they were destined at a level of probability of 90% to die of kuru, the survival of these people demanded that each female have 2 or more pregnancies before dying. Such a pattern of early marriage constituted conscious adaptation to a disease that cut short the breeding life of their women.

The only source of the infection was cannibalism, cessation of which is causing kuru to disappear. No one born since ritualistic cannibalism stopped in his area has ever developed the disease. All childhood kuru has thus been eliminated. We expect kuru will have killed only about 20 persons in 1976, and the disease will be gone from the world in about 10 years.

Epidemiologic studies of kuru gave no evidence for contagiousness. No outsiders to the Fore region ever developed kuru after residence in the area. The disease has, however, developed in Fore natives years after they left the region for schooling or work. We have had more than 100 such cases. Some developed the disease while residing at distant places, such as New Britain or the Sepik, as long as 10 years after leaving their homeland. Such long latency was regarded as unique in medicine, and it told us much about the incubation period and also ruled out the possibility that nutritional deficiency or some toxic factor might be involved. That the disease is not transmitted transplacentally to children is indicated by the finding that, since cannibalism has ceased, of more than 300 children born to mothers who suffered from kuru during pregnancy, not one child has developed the disease. Thus, kuru-affected mothers are not transmitting childhood or adolescent kuru to their infants.

Not all new cases reported to us as kuru were evaluated as true kuru. In one village of about 200 people, each of 11 women said she had kuru. We convinced ourselves by neurological examination that only six of the group had the disease. The other five, we were reasonably sure, were hysterically mimicking kuru or, for social reasons, decided they were going to die of kuru. Those mimicking kuru tended to spring into the late stages of the disease overnight and, after a series of socially rewarding ceremonies of divination and cure, they sprung suddenly out of their diseased condition to an overnight cure. The people themselves recognized such behavior as a special type of the disease. We are sure those patients were showing hysterical mimicry. Few patients were in that group, but that phenomenon, and only that phenomenon, gave rise to published accounts of apparent recovery. No case of kuru with hard neurological symptoms of organic brain disease has recovered.

Creutzfeldt-Jakob Disease

As early as 1959, Klatzo et al. (11) pointed out the similarity between CJD, a progressive fatal CNS dementia-producing disease of middle life, and kuru. In the United States and Europe, CJD was identified in earlier medical literature as a sporadic dementia of the spongiform encephalopathy type. It was known by many other names, most of which are eponymic, a circumstance that is not unusual for degenerative disease.

The disease was the first presenile dementia of man demonstrated to be due to a transmissible agent (12). CJD appeared to be genetically transmitted, assuming a familial pattern of autosomal dominant inheritance in more than 10% of the cases, with cases sometimes occurring over several generations. However, if brain or liver tissue obtained by biopsy or early autopsy from these familial cases of CJD is inoculated into susceptible hosts, it produces disease.

The clinical picture of CJD is very variable. The mean age of onset is approximately 55 years; it occurs rarely before age 35 and the incidence in men and women is about the same (13). CJD often starts with vague symptoms; some victims having predominantly lower motor neuron signs, some cerebellar deficits, and others severe visual disturbances, but all develop progressive dementia. In addition, the patient may have many other CNS disturbances, including psychosis, extrapyramidal disturbances, myoclonus, characteristic (though not specific) high-voltage slow paroxysmal bursts on EEG, and evidence of widespread cerebral dysfunction. Death usually occurs within two years of onset, often in less than one year.

CJD hitherto was believed to occur with exceedingly low frequency. Now, however, it is recognized to be of worldwide occurrence, with an annual incidence of about 1 to 3 cases per million population. Surveys show that CJD kills at least 200 Americans each year. In the United States, the incidence may even be as high as 2 to 3 per million, as it was found to be in Israel and in the academic centers of the United States, Canada and the United Kingdom, Finland, France, Switzerland, and Chile. CJD has also been found in Japan, the Soviet Union, Norway, Sweden, Holland, Germany, Italy, Peru, Colombia, Brazil, Argentina, Australia, and Papua New Guinea. Its incidence is especially high in Jews of Libyan origin (14), in whom the incidence is more than 30-fold higher than in Jews of European origin. We have even documented the disease in one of the stone-age Melanesian New Guineans far from the kuru area, in the Chimbu area. Apparently, CJD can be found wherever it has been carefully sought.

LABORATORY FINDINGS

Although the spinal fluid proteins may be increased in the subacute spongiform encephalopathies, no consistent quantitative biochemical abnormalities are seen, even by electrophoretic study. We usually found no elevated cell counts, although many hundreds of lumbar punctures and cell counts have been done on kuru patients and CSF proteins were not elevated. These findings are certainly unusual in an infectious process. In CJD pneumoencephalography may be normal or may show ventricular enlargement and cerebral or cerebellar cortical atrophy.

The agents of the spongiform encephalopathies appear to be totally lacking in antigenicity for their hosts. No other infectious disease of man shows so little influence, if any, on the immune system. All attempts to detect serum antibodies and delayed hypersensitivity to the agent(s) in man or experimental animals have been unsuccessful.

To date, even in animals hyperimmunized with preparations that have titers of more than 10^{11} infectious units of scrapie per inoculum (10^{12} or 10^{13} infectious units per gram of hamster brain tissue) we have been unable to demonstrate any antigenic activity. Schedules for hyperimmunization, such as footpad inoculation in guinea pigs, the use of avian hosts, and peritoneal macrophages that produce antibodies for other agents of low antigenicity (e.g., oncornaviruses), have been unsuccessful with the unconventional agent. No antigen/antibody complexes have been found on the glomerular membranes or in the choroid plexus to explain the absence of antibodies (15).

The basic immune mechanisms appear to be intact. "T" and "B" cell functions *in vivo* or *in vitro* during any stage of incubation, even late in the disease, are not affected. Responses to homologous and heterologous antigens are within normal limits.

Attempts to immunize animals with high-titered preparations of the scrapie agent have failed. Immunosuppression has had no effect on the incubation period or the course of disease, whether administered early or late in the disease or even before inoculation.

Repeated treatment with concentrated interferon and with interferon inducers did not similarly protect mice from scrapie. The finding that amantadine improved the clinical status of patients with CJD (16) remains unconfirmed.

PATHOLOGY

On postmortem examination the brains of patients with subacute spongiform encephalopathies may appear to be grossly normal, or they may show some cerebellar atrophy in kuru and cerebral cortical atrophy in CJD. By light microscopy the gray matter shows neuronal loss, gliosis, and vacuolation of the neuropil, in the absence of leukocytic infiltrates or other evidence of an immune response; there is no perivascular cuffing of the kind seen in other virus infections of the CNS.

In kuru, no gross brain lesions may be evident except for atrophy of the cerebellar vermis. The cerebrum is involved minimally or not at all. Kuru affects mostly the midline structures and the cerebellum. Microscopic changes are widespread and are characterized by proliferation and hypertrophy of the astrocytes and sometimes the microglia throughout the brain, diffuse neuronal loss that is most severe in the cerebellum and in its afferent and efferent connections, mild spongiform change of the gray matter, and minimal loss of myelinated axons. The latter, although not readily seen in sections stained for myelin, can be detected by demonstrating the presence of neutral fat. In addition, most brains have plaques that contain amyloid. Intracytoplasmic vacuolation may be observed in the large neurones of the corpus striatum. Electronmicroscopic studies of brain tissue show that the spongiform change is associated with progressive clearing and eventual vacuolation of neuronal cytoplasm, particularly in presynaptic and postsynaptic processes of the neurones, finally with cell destruction (17). These changes occur to a lesser extent in astrocytes and oligodendrocytes. Ruptured plasma membranes and curled fragments of membranes are seen within the vacuoles. It appears that astrocytes react by proliferation and hypertrophy, whereas nerve cells are lost.

Neuropathologists who observed the lesions dismissed hypersensitivity or infection as possible causes because of the absence of any of the pathological hallmarks of previously known infections or hypersensitive reactions in the CNS. Rarely, a blood vessel may be cuffed with mononuclear cells, the pathological process being otherwise devoid of any evidence of inflammatory response. Because more than half of the Purkinje cells and more than three fourths of the granule cells may be

242

gone at the time of death, these patients in effect suffer cerebellar extirpation. However, the cerebrum remains reasonably intact in kuru, and the spinal cord usually shows no lesions.

In CJD, the histopathological changes throughout the gray matter of the brain are most severe in the cerebral cortex and striatum but also occur in the thalamus, cerebellum, and brain stem. The triad of vacuolation (spongiform change), neuronal loss, and astrocytic hyperplasia is almost identical to that seen in kuru. At an ultrastructural level the vacuolating change is identical to that seen in kuru.

EXPERIMENTAL TRANSMISSION OF SPONGIFORM ENCEPHALOPATHIES *IN VITRO* AND *IN VIVO*

All four subacute spongiform encephalopathies have been transmitted to a variety and an overlapping range of experimental animals (Tables 13.4 and 13.5) using bacterial-free filtrates of infected tissues. We can reisolate the agent in high titer from the tissues of every affected host, whether human or animal.

TABLE 13.4. *Species of Laboratory Animals Susceptible to the Subacute Spongiform Viral Encephalopathies*

Kuru	
Chimpanzee	Mink
New World monkeys	Ferret
Old World monkeys	
Creutzfeldt-Jakob disease	
Chimpanzee	Cat
New World monkeys	Guinea pig
Old World monkeys	Mouse
Hamster	Ferret
Scrapie	
New World monkeys	Mouse
Old World monkeys	Rat
Sheep	Hamster
Goat	Gerbil
Mink	Vole
Transmissible mink encephalopathy	
New World monkeys	Ferret
Old World monkeys	Raccoon
Sheep	Skunk
Goat	Hamster
Mink	

TABLE 13.5. *Approximate Incubation Periods and Duration of Diseases Caused by Unconventional Viruses*

	Kuru	CJD	Scrapie	TME
Natural disease				
Incubation	5–30 years	Unknown	2–5 years	Unknown
Duration	3–12 months	3–36 months	2–6 months	3–8 weeks
Experimental disease in primates				
Incubation	10–101 months	10–71 months	14–65 months	11–33 months
Duration	1–11 months	1–5 months	2 months	1 month

Kuru was first transmitted to and passaged in chimpanzees (18,19), which developed a disease resembling kuru in man. Later the disease was transmitted to New World and Old World monkeys. The kuru virus passed in squirrel monkeys may cause brain and cord lesions as well after variable incubation periods ranging from 10 months to several years. Kuru may kill squirrel monkeys with a disease as acute as that seen in rabies, that is, in from one to three days after onset of the disease. However, the same strain of kuru virus inoculated into spider monkeys and chimpanzees caused disease that paralleled that in humans, a slowly evolving syndrome with minimal dementia, extreme cerebellar ataxia, and an incubation period of two to three years. Clear-cut neurological signs are absent during the early months after inoculation. Such an enormously prolonged incubation period was previously unknown in virology, except perhaps in tumor virology. The affected animals regularly show personality changes, for example, voluntary withdrawal from running to pick up food (although, if forced to do so, they move rather well). Interestingly, various species of New World monkeys inoculated with kuru developed ataxia in their "fifth limb," the prehensile tail.

Isolation of the kuru agent in experimental animals led to the demonstration that CJD is also caused by an infectious filterable agent (20), although the natural mode of transmission is unknown. Intracerebral or peripheral inoculations of suspensions of brain, liver, kidney, lymph node, lung, cerebrospinal fluids, or brain tissue grown in culture have produced experimental disease closely resembling human CJD clinically and pathologically but differing from kuru, after incubation periods of months to several years, depending on the animal recipient. Transmission of CJD was accomplished in less than a year in the chimpanzee and that of kuru in one and one half to three years, both representing a shorter time than was possible in other primates. One chimpanzee at five months of age already showed the burst suppression and high-voltage waves on electroencephalographic (EEG) recordings from implanted electrodes. Even as early as three months after challenge, a computer showed the EEG wave changes clearly. Interestingly, the animal remained clinically and psychologically intact for the ensuring year, and not until 13 months after inoculation did we note any psychological changes. Two months later, at age 15 months, the animal was in advanced stages of the disease.

Cats are now known to be susceptible to CJD, the incubation period being about three years. The fur undergoes remarkable deterioration because the animals stop grooming; later, ataxia, circling, and myoclonic jerks occur. Dozens of cats are currently under study, having been inoculated with either kuru or CJD agents. The first EEG evidence of early CNS derangement occurs before veterinarians or neurologists can detect any definitive signs.

Scrapie and TME have even wider host ranges. Scrapie given experimentally to mink is modified in one passage to acquire the properties of TME. Therefore, the thesis is tenable that mink encephalopathy results from the passage of an agent to mink from scrapie-infected sheep. This passage sequence occurred on two mink ranches in the United States where carcasses of sheep with scrapie were used as feed, which we have good reason to believe was the source of infection. Obviously, investigation of scrapie as a potential agent of human disease is essential. Scrapie occurs in most states of the Union, and affected meat probably finds its way to market because, though infected, it passes the usual routine inspection. Scrapie is transmissible by feeding to mink, and the disease is transmitted among mink by biting. One form of linear transmission is by contamination of the ground and the consumption of placental membranes and amnions from animals infected with scrapie. When introduced into primates and mink, the scrapie agent, although serially transmissible in its new hosts, appears to have lost its ability to replicate readily in mice.

Tissues from animals inoculated with kuru and serially examined for early lesions by light and electron microscopy showed fusion of neurones with astroglial or oligodendroglial cells, which presumably occurred *in vivo* soon after inoculation. This fusion was the same kind observed by electron microscopic examination of tissues of some animals having scrapie or CJD (21). Fusion was coupled with stimulation of mitosis in neurones that do not normally divide. This undoubtedly explains the odd finding, repeatedly reported at autopsy, of binucleate nuerones in human victims of CJD or kuru. We now find that these strange mitoses, even in Purkinje cells, are an early phenomenon in the spongiform encephalopathies. No concomitant astrogliosis occurs until many months later. What this finding means is unknown.

These agents can be sustained *in vitro* in cultures of infected brain tissue (22), which can be trypsinized; in a few instances, they have been cloned and carried as transformed cell lines for many subculture generations. Small amounts of inocula of cultured material have induced the respective disease in chimpanzees. Cultures of scrapie have been maintained from 18 months to 2 years, retaining their infectivity. Although the agents replicate in primary cell cultures *in vitro,* the problem of transferring infectivity to other cell lines has been almost insurmountable. Many dozens of continuous cell lines have been assayed. Recently, in England, with the use of lysolecithin and chelating agents, a line of L cells was made to carry the scrapie agent. We have also succeeded in establishing the scrapie virus in cell lines derived from mouse or sheep brain cells transformed by SV40.

THE VIRUS(ES)

Although the clinical and histopathological pictures of the spongiform encephalopathies differ in their natural and experimental hosts, the general similarity of the four diseases, especially in histopathological lesions and their partially coextensive host range, suggests that the causative agents may be variants of one original virus, somewhat altered in biologic properties.

Despite the enormous amount of work done to learn the biochemical nature of the infective virus, there is still no clearly discernible method whereby the agents can be recognized physically or chemically. They can be detected only by their infectivity and their unique stability.

The physical properties of the four agents, which are indistinguishable, suggest that they must have a structure that differs greatly from those of known microbial pathogens of animals. The kuru agent passes through millipore filters having a pore size of 220 nm but not through ones having a 100 nm or smaller average pore diameter (23). The size of the scrapie agent has been estimated by membrane filtration to be between 50 and 27 nm (24) and, by gel filtration, at least 5×10^6 daltons (25). Membrane filtration of the agent of TME indicated that it also passed through pores of 50 nm average diameter (26).

Particles of such size are small, but they should be visible by electron microscopy if present in sufficient quantity. No one has seen a virion or virionlike structure in partially purified preparations containing more than 10^6 infectious doses per milliliter (lethal doses for mice) of the scrapie agent. However, vesicular structures of about the right size have been described in intact brain cells of animals with scrapie (21,27,28), in chimpanzees with CJD (17), and in brain cells of patients with CJD (29,30). Whether any of these structures is a particle of the infectious agent is not clear.

The agent or agents differ from all other known microbial agents in their ability to withstand adverse conditions, including many virus-destroying solutions. The scrapie agent is highly resistant to ionizing radiation, which suggests a much smaller "target" size than that estimated by filtration (perhaps only 2×10^5 daltons). It is also more resistant to ultraviolet light than any other known pathogen (31,32).

The agents of scrapie, TME, and kuru are highly resistant to heat (24). Although there is rapid loss of infectivity at temperatures above 85°C, some infectivity remains even after 30 minutes of boiling. The agents are resistant to nucleases and proteases. The scrapie agent is also relatively resistant to treatment with 10% formalin, to a pH range of 2.5 to 10.5, and to betapropiolactone and ethylene diamine tetracetic acid (EDTA) (all of which inactivate conventional viruses), but it is sensitive to treatment with periodate, 95% phenol, 6–8 M urea, 2-chloroethanol, and some detergents and lipid solvents (33). The agent of TME behaves similarly (34). When homogenates of scrapie-infected tissues are fractionated, infectivity appears to be associated with preparations rich in fragments of normal membranes that are probably plasma membranes and perhaps endoplasmic reticulum (35). The infectious preparations of the scrapie agent always contain fragments of membranes of the host cell, and the infectivity of scrapie is destroyed by reagents that dissociate membrane structures as well as by several reagents that inactivate other cell components, all suggesting a close association with, and location in, host-cell membranes. The resistance of the agent to ionizing radiation suggests that its replicating portion is probably exceedingly small (much smaller than the overall size as indicated by filtration) and its resistance to ultraviolet radiation with an inactivation spectrum peak at 236 nm, rather than at 260 nm, has been interpreted as indicating that the virus contains no nucleic acid (36) or that, if the replicating moiety is nucleic acid, it is probably much smaller than the nucleic acids of conventional viruses. It is probably protected from inactivation by close binding to membrane structures.

An alternative coding mechanism, such as self-replication of abnormal membranes, has been postulated (37). However, it may be premature to dismiss nucleic acid as the genetic material of these agents. For example, recent studies have shown atypical agents to be associated with classical plant diseases such as potato spindle tuber, chrysanthemum stunt, citrus exorcortis diseases, and Cadang-Cadang disease of coconuts. These viruslike pathogens that appear to be naked molecules of RNA about 120,000 daltons (38) contain no core or coat protein. These plant "viroids" resemble the scrapie agent in that they are not visible by electron microscopy in tissue sections, and are extremely resistant to inactivation by UV irradiation—and also in their failure to direct the synthesis of any antigens of their own.

Because viroids differ from the scrapie agent in their resistance to treatment with phenol, urea, and lipid solvent and are thought to be associated with nuclear RNA rather than cytoplasmic membrane components (38), it is premature to conclude that the scrapie agent is a viroid. However, the plant viroids serve to demonstrate that a self-replicating nucleic acid of small size can still carry sufficient information to elicit disease and that such agents have high UV resistance and contain no proteins of their own. Perhaps the agents of the spongiform encephalopathies contain tiny pieces of nucleic acid tightly bound to fragments of plasma membranes.

Obviously a helper virus is not involved in the transmission of scrapie, because formaldehyde or heat or UV treatment would block transmission of a second or third helper. Finally, the absence of any viral protein distinct from the host, even in highly infectious preparations, is extremely puzzling.

Hitherto, the principal basis of classification of viruses has rested on serologic specificity. Because of their incredible properties, perhaps this peculiar group of pathogens should not be called viruses. Obviously they differ from all the other viruses as listed in the latest taxonomy (see Chapter 2). We have deferred renaming them until we know more about their structure.

IMPLICATIONS FOR OTHER DISEASES

Once we knew that a degenerative disease could be caused by a virus after incubation of many years, as we now know to be true for kuru and CJD, obviously we had to reconsider the pathogenesis of multiple sclerosis, amyotrophic lateral sclerosis, Parkinson's disease, and, particularly, Alzheimer and senile dementias, and other familial dementias.

PAS-positive, silver-staining, amyloid-containing plaques, like the plaques in senile brains, appear in 60% of kuru patients, including the young children with the disease (39). Several investigators have reporte 1 finding amyloid plaques, like those of kuru in patients with CJD (40,41). Some cases of CJD are difficult to distinguish from Alzheimer's disease, Pick's disease, senile dementia, or dementia parkinsonism, either clinically or pathologically. With brain tissues from such cases, which contain plaques and neurofibrillary tangles typical of Alzheimer's disease, and with the clinical features of this disease, a subacute spongiform encephalopathy has been transmitted to squirrel, capuchin, and spider monkeys. On transmission to primates, a clear-cut spongiform encephalopathy like that in CJD developed, and no plaques or tangles were seen. Brain tissue from many sporadic cases of Alzheimer's disease similarly inoculated into primates has *not* caused disease in these animals.

We are not claiming that Alzheimer's disease is a slow virus disease, and we are not contending that the classical sporadic Alzheimer's disease is related to CJD. On the other hand, having found two cases of Alzheimer's disease that contained the virus of transmissible virus dementia, we must suspect that the apparently genetic, familial form of the disease may be caused by a virus. To determine whether the familial form of Alzheimer's disease has the same origin as familial cases of CJD is now our problem.

The concomitant extreme astrocytosis demonstrated by Holzer's stain or by staining with silver or gold impregnation, and usually found in the subacute spongiform encephalopathies, is as extreme as we have ever seen in neuropathology. We find it in both cerebral and cerebellar tissue in all cases of the natural infections. Such astrogliosis (but much milder) and occasional amyloid plaques, are not uncommon in aged brains. There are many similarities between the neuropathology of kuru and CJD and that seen in senile change. These similarities give rise to the speculation that some of the changes in the aging brain may result from activation of latent viral agents with aging (42).

We are also finding that some chronic neurological diseases are not infections caused by unconventional viruses but result from classical viruses behaving in a defective manner, with asynchronous synthesis of subunits and failure of assembly (Table 13.6). The patients die without the whole genome having been expressed in the cells. But in these diseases, there is usually an inflammatory reaction consisting of mononuclear cell infiltration and perivascular cuffing.

Measles virus, or a close variant of it, has been associated causally with subacute sclerosing panencephalitis. Pseudomyxoviral elements were observed by electron microscopy in brain tissue from patients who had high titers of antibody to measles virus in both serum and cerebrospinal fluid. The evidence suggests that in this disease there is defective expression of the viral genome and asynchronous production of the subunits. Through heterokaryon formation by cocultivation of brain tissue with susceptible cells, the appearance of complete antigen

TABLE 13.6. *Slow Infections of Man Caused by Conventional Viruses*

Disease	Virus
Subacute post-measles leukoencephalitis	Paramyxovirus—defective measles
Subacute sclerosing panencephalitis (SSPE)	Paramyxovirus—defective measles
Subacute encephalitis	Herpetovirus—herpes simplex
	Adenovirus—adeno-types 7 and 32
Progressive congenital rubella	Togavirus—rubella
Progressive panencephalitis as a late sequela following congenital rubella	Togavirus—defective rubella
Progressive multifocal leukoencephalopathy (PML)	Papovavirus—JC; SV40
Cytomegalovirus brain infection	Herpetovirus—cytomegalovirus
Epilepsia partialis continua (Kozhevnikov epilepsy) and progressive bulbar palsy in USSR	Togavirus—RSSE and other tick-bone encephalitis viruses
Chronic meningoencephalitis in immuno-deficient patients	Picornaviruses—poliomyelitis ECHO virus
Crohn disease	Unclassified—RNA virus
Homologous serum jaundice	Unclassified—hepatitis B, Dane particle
Infectious hepatitis	Parvovirus—hepatitis A
	Unclassified—hepatitis B, Dane particle
	Unclassified—hepatitis C

is induced. After the antigens appear, further genome expression results in budding of the whole infectious unit from the cell membrane. This slow resurrection of the agent has been possible in no more than 1 in 10 specimens. There is no proof that in the other 9, a whole genome is even there to be expressed.

Epilepsia partialis continua, or Kozhevnikov's epilepsy, has been associated with a chronic focal infection of the brain, Russian spring-summer encephalitis (RSSE), caused by the tick-borne virus that produces low-grade disease lasting for years.

Progressive multifocal leukoencephalopathy (PML) is a rare demyelinating disease that invariably occurs in patients with leukemia, lymphoma, and other immunosuppressive diseases or genetic immune deficiencies (see Chapter 11). Papovaviruses, such as JCV, have been isolated from brains of PML patients. JCV does not replicate in the oligodendrocytes defectively, as might be expected if there were asynchronous production of subunit protein or failure of assembly. In fact, when fresh brain tissues of patients are examined by measuring infectivity as a function of virion count, a higher proportion of the virions are infectious than in virus from tissue culture systems. It appears, therefore, that the brain in the immunosuppressed patient supports synthesis of complete infectious virus units.

One wonders, therefore, whether multiple sclerosis, lupus erythematosus, and a whole gamut of degenerative disease might have a similar pathogenesis. To prove that easily made hypothesis, the technology for identifying virus subunits in tissue must advance enormously and the work must involve probes using DNA homology.

The lesson gained is this: Do not make any general rules about slow virus diseases and their pathogenesis. Slow infections with conventional viruses may involve defective virus replication or be caused by complete synthesis of fully assembled and infectious virions. Conventional viruses cause most of the slow infections, but one wonders how many other diseases fit into the list of those caused by unconventional viruses and whether any really important diseases of high prevalence belong there.

A PRAGMATIC PERSPECTIVE

Although many basic questions about the pathogenesis of the subacute spongiform encephalopathies remain unanswered, many aspects of prevention and control also need to be resolved.

That scrapie virus causes a disease indistinguishable from CJD in several species of New World and Old World monkeys (although, admittedly not yet in chimpanzees) suggests the possibility of it as the source of some CJD in man. The existence in butcher shops of the pathogenic agent of scrapie in sheep tissues poses a possible potential source of human infection. We wonder whether kitchen or butchery accidents involving inoculation of scrapie-infected sheep tissue by rubbing the eyes, picking the nose, or through breaks in the skin are contributory to the disease in man (2,4).

CJD has been transmitted by corneal transplantation. A year and a half after receiving a cornea from a donor who had the disease, the recipient developed it and died (43). From the brain of the corneal recipient, preserved for nine months in formaldehyde in a pathologist's laboratory, we isolated the agent (44). The potential hazard of even formaldehyde-fixed CJD tissue to the pathologist is obvious. Whether other corneas were infected by contact with the instruments used in handling the infected cornea in the corneal bank is unknown. Obviously, the potential psychosomatic problems of attempted follow-up study of other corneal transplant patients, and the legal implications, are serious.

Tissue implants and blood transfusion might also be a source of infection. We have seen four professional blood donors who developed the disease, one of whom, during the onset of dementia, continued to give units of blood for use in a pediatric unit. The potential occupational hazard for physicians and pathologists should be noted. The case may be cited of a professor of neurosurgery who died of CJD at age 54 with a high titer of virus in his peripheral tissues and brain (45). His may have been an unusual accidental infection. Alternatively, his long-standing enthusiasm for doing autopsies and studying pathological specimens may have exposed him to the virus.

The potential occupational hazards for pathologists and neurosurgeons are obvious. That neurosurgery itself may introduce the agent has been suggested (3,45). Matthews (46) reported that 7% of his patients having the disease had had neurosurgery for other causes in the previous five years. In our own series, 5% of the patients had had previous brain surgery (3). We thought at first that this was good evidence that anoxia, anesthesia, or other surgical factors might have precipitated active infection by a latent virion already present. On the other hand, Matthews and we have suggested that an inadequately sterilized neurosurgical instrument, such as the brain microscope, might have introduced the agent. This unfortunate possibility has now actually occurred

(47). Many cases of CJD may thus be iatrogenic. We have recently suggested precautions in the medical care and in handling medical materials from patients with CJD (48).

Much work remains to be done to further characterize the viruses causing spongiform encephalopathies. Undoubtedly the effort will join together many disciplines and investigators to achieve that purpose.

DISCUSSION

Question: Is the increased incidence of CJD due to more intense interest in the search for the disease or has the incidence really increased?

Dr. Gajdusek: It appears that the increased reporting of CJD is most certainly due to increased diagnosis and increased awareness by pathologists and neurologists. Many neurologists visiting Israel were unsure that it had ever occurred there until Kahana, Alter, and their group studied the records of hospitals, and finally found that the incidence was the same as elsewhere in the world—one to two deaths per million population per annum in the whole country. In any area where physicians have really hunted for the disease, they have been rather surprised to find such a high incidence. We estimate that about 200 cases per annum occur in the United States, but there probably are more. Patients with the disease are not likely to be admitted to academic hospitals or, if admitted, they do not remain there long; older persons with presenile dementia have been in the past of only remote interest for academic neurology. Patients with dementia are usually transferred to psychiatric units or often to nursing homes. CJD probably accounts for one death in every 5,000 to 10,000 deaths in persons over 20 years of age.

Question: A paradox seems to exist in that measles is being implicated in the pathogenesis of multiple sclerosis. Yet, in some places in the world, the incidence of measles is higher but that of multiple sclerosis is much lower, whereas in other areas the incidences are reversed. Could these different incidences be due to diet, environment, or genetic influences, and what other factors might cause this sort of paradox?

Dr. Gajdusek: The whole multiple sclerosis story is rather complicated.

This paper has been prepared from a transcript of a lecture by Dr. Gajdusek and has been approved.

You all are aware that in the tropical regions of the world, multiple sclerosis occurs in a much lower incidence than it does in temperate areas. Persons who have migrated to New Zealand, Australia, the United States, and other places after adolescence bring with them the incidence of the multiple sclerosis of their place of origin. This retained incidence occurs as long as the migrants left their homelands after adolescence, and, therefore, it seems as though some childhood insult determines how likely one is to get multiple sclerosis. The childhood insult may well be an infectious disease. Not enough is known about the association between measles and multiple sclerosis, and further study is needed to determine what type of measles infection *might* lead to multiple sclerosis.

Even in SSPE, most of the patients have had an atypical type of measles in the first two years of life, in fact, a significant number in the first year. Measles in such circumstances is rather atypical in that it occurs during the period of incomplete maturation of the immune system and during partial protection with maternal antibody. Whether measles causes some cases of multiple sclerosis is not known. We know for sure that measles does not cause all cases of multiple sclerosis. One case of multiple sclerosis brought to my attention in the United States and one case in Europe occurred in measles virgins who had no signs of measles antibody, and in one of these patients, a challenge with the killed measles antigen produced a primary, not an anamnestic, response. The multiple sclerosis in this patient, therefore, could not have had anything to do with a previous measles infection. However, among patients with the Devic syndrome (neuromyelitis optica, a rather acute form of multiple sclerosis having more inflammation than the common type), Norrby in the Karolinska Institutet and others have found that about 50% of such patients have very high, abnormal titers of oligoclonal IgG measles antibodies in the spinal fluid.

Question: Is any spongiform encephalopathy induced by filtrates of brain from normal or senile brains?

Dr. Gajdusek: We haven't been using many inocula from normal brain, but inocula from brains of senile patients with senile dementia in their 80s and 90s have gone into primates for long incubations—and nothing has happened to date. The control of the work is, of course, inherent in the inoculation of several hundred brain biopsy and autopsy specimens from patients who had multiple sclerosis, Parkinson's disease, amyotrophic lateral sclerosis, and Parkinsonism-dementia complex from Guam and other chronic CNS diseases, none of which has produced any disease in the inoculated animals. Some of the animals have now been maintained for well over 10 years. Consequently, we have no evidence that the agent is ubiquitous. However, the occurrence of insults that may "turn on" the infection, like trauma and cerebrovascular accidents in a few cases of the Creutzfeldt-Jakob syndrome, confuse the issue. Perhaps we must allow that, if the agent is ubiquitous, its replicating simply isn't "turned on" and, therefore, we can inoculate primates without producing disease.

Question: Have any studies been done on the relatives of Creutzfeldt-Jakob patients, relatives who die of other causes?

Dr. Gajdusek: The epidemiology of the 36 cases of known familial Creutzfeldt-Jakob disease throughout the world is just under way. We don't have much information about any significant related disease. The familial cases of CJD have already been selected for epidemiologic study because they include two or more cases in near relatives, and we thought that would be a much better area for epidemiologic sleuthing than the remaining 80% of the cases that are totally sporadic.

REFERENCES

1. Gibbs CJ Jr, Gajdusek DC: Infection as the etiology of spongiform encephalopathy (Creutzfeldt-Jakob disease). *Science* 165:1023–1025, 1969.
2. Gajdusek DC, Gibbs CJ Jr: Slow virus infections of the nervous system and the Laboratories of Slow, Latent and Temperate Virus Infections, in Chase TN (ed): *The Clinical Neurosciences.* Vol 2: Tower DB (ed): *The Nervous System,* New York, Raven Press, 1975, pp 113–135.
3. Traub R, Gajdusek DC, Gibbs CJ Jr: Transmissible virus dementias. The relation of transmissible spongiform encephalopathy to Creutzfeldt-Jakob disease, in Kinsbourne M, Smith L (eds): *Aging, Dementia and Cerebral Function.* Flushing, NY, Spectrum Publishing Inc., 1977, pp 91–146.
4. Gajdusek DC: Unconventional viruses and the origin and disappearance of kuru. *Science* 197:943–960, 1977. Also in *Les Prix Novel 1976,* pp 167–216.
5. Sigurdsson B: Observations on three slow infections of sheep. *Br Vet J* 110:7–9, 255–270, 307–322, 341–354, 1954.
6. Gordon WS: Variation in susceptibility of sheep to scrapie and genetic

implications. *Report of Scrapie Seminar*, U.S. Department of Agriculture, ARS 91-53 (May) 1964, pp 8–18.

7. Cuillé J, Chelle PL: Pathologie animale—la maladie dite tremblante du mouton est-elle inoculable? *CR Acad Sci (D) (Paris)* 203:1552–1554, 1936.

8. Hartsough GR, Burger D: Encephalopathy in mink: I. Epizoological and clinical observations. *J Infect Dis* 155:387–392, 1965.

9. Gajdusek DC, Zigas V: Degenerative disease of the central nervous system in New Guinea: The epidemic occurrence of "kuru" in the native population. *N Engl J Med* 257 (30):974–978, 1957.

10. Gajdusek DC, Alpers M: Genetic studies in relation to kuru: I. Cultural, historical, and demographic background. *Am J Hum Genet* 24:S1–S38, 1972.

11. Klatzo I, Gajdusek DC, Zigas V: Pathology of kuru. *Lab Invest* 8:799–847, 1959.

12. Gibbs CJ Jr, Gajdusek DC, Asher DM, et al: Creutzfeldt-Jakob disease (subacute spongiform encephalopathy): Transmission to the chimpanzee. *Science* 161:388–389, 1968.

13. Roos R, Gajdusek DC, Gibbs CJ Jr: The clinical characteristics of transmissible Creutzfeldt-Jakob disease. *Brain* 96:1–20, 1973.

14. Kahana E, Alter M, Braham J: Creutzfeldt-Jakob disease: Focus among Libyan Jews in Israel. *Science* 183:90–91, 1974.

15. Porter DD, Porter HG, Cox NA: Failure to demonstrate a humoral immune response to scrapie infection in mice. *J Immunol* 111:1407–1410, 1973.

16. Sanders WL, Dunn TL: Creutzfeldt-Jakob disease treated with amantadine: A report of two cases. *J Neurol Neurosurg Psychiatry* 36:581–584, 1973.

17. Lampert PW, Gajdusek DC, Gibbs, CJ Jr: Experimental spongiform encephalopathy (Creutzfeldt-Jakob disease) in chimpanzees: Electron microscopic studies. *J Neuropathol Exp Neurol* 30:20–32, 1971.

18. Gajdusek DC, Gibbs CJ Jr, Alpers M: Experimental transmission of a kuru-like syndrome to chimpanzees. *Nature* 209:794–796, 1966.

19. Gajdusek DC, Gibbs CJ Jr, Alpers M: Transmission and passage of experimental "kuru" to chimpanzees. *Science* 155:212–214, 1967.

20. Gibbs CJ Jr, Gajdusek DC, Alpers MP: Attempts to transmit subacute and chronic neurological diseases to animals: With a progress report on the experimental transmission of kuru to chimpanzees, in Burdzy K, Kallós P (eds): *Pathogenesis and Etiology of Demyelinating Diseases*. New York, S. Karger, 1969.

21. Lampert P, Hooks J, Gibbs CJ Jr, et al: Altered plasma membranes in experimental scrapie. *Acta Neuropathol* 19:81–93, 1971.

22. Gajdusek DC, Gibbs CJ Jr, Rogers NG, et al: Persistence of viruses of kuru and Creutzfeldt-Jakob disease in tissue culture of brain cells. *Nature* 235:104–105, 1972.

23. Gibbs CJ Jr, Gajdusek DC: Characterization and nature of viruses causing subacute spongiform encephalopathies, in *Proc VIth Int Cong Neuropath*. Paris, Masson et Cie, editeurs, 1970, pp 779–801.

24. Gibbs CJ Jr, Gajdusek DC: Transmission and characterization of the agents of spongiform virus encephalopathies: Kuru, Creutzfeldt-Jakob disease, scrapie and mink encephalopathy. *Res Publ Assoc Res Nerv Ment Dis* 49:383–410, 1971.

25. Kimberlin RH, Millson CF, Hunter GD: An experimental examination of the scrapie agent in cell membrane mixtures: III. Studies of operational size. *J Comp Pathol* 81:383–391, 1971.

26. Marsh RF, Hanson RP: Physical and chemical properties of the transmissible mink encephalopathy agent. *J Virol* 3:176–180, 1969.

27. David-Ferreira JF, David-Ferreira KL, Gibbs CJ Jr, et al: Scrapie in mice: Ultrastructural observations in the cerebral cortex. *Proc Soc Exp Biol Med* 127:313–320, 1968.

28. Narang HK, Shenton B, Giorgi PP, et al: Scrapie agent and neurones. *Nature* 240:106–107, 1972.

29. Bots GthAM, deMan JCH, Verjaal A: Virus-like particles in brain tissue from two patients with Creutzfeldt-Jakob disease. *Acta Neuropathol* 18:267–270, 1971.

30. Sever JL, Horta-Barbosa L, Vernon ML, et al: Creutzfeldt-Jakob disease: Virus-like particles in brain biopsies and tissue cultures, in *Proc VIth Internat Cong Neuropathol*. Paris, Masson et Cie, editeurs, 1970, pp 931–932.

31. Gajdusek DC, Gibbs CJ Jr: Slow, latent and temperate virus infections of the nervous system. *Res Publ Assoc Res Nerv Ment Dis* 44:254–280, 1968.

32. Latarjet R, Muel B, Haig DA, et al: Inactivation of the scrapie agent by near monochromatic ultraviolet light. *Nature* 227:1341–1343, 1970.

33. Hunter GD: Scrapie, *Prog Med Virol* 18:289–306, 1974.

34. Marsh RF, Semancik JS, Medappa KC, et al: Scrapie and transmissible mink encephalopathy: Search for infectious nucleic acid. *J Virol* 13:993–996, 1974.

35. Millson GC, Hunter GD, Kimberly RH: An experimental examination of the scrapie agent in cell membrane mixtures: II. The association of scrapie activity with membrane fractions. *Comp Pathol* 81:255–265, 1971.

36. Alper T, Cramp WA, Haig DA, et al: Does the agent of scrapie replicate without nucleic acid? *Nature* 214:764–766, 1967.

37. Gibbons RA, Hunter GD: Nature of the scrapie agent. *Nature* 215:1041–1043, 1967.

38. Diener TO: Viroids: The smallest known agents of infectious disease. *Ann Rev Microbiol* 28:23–39, 1974.

39. Beck E, Daniel PM, Gajdusek DC, Gibbs CJ Jr: Similarities and differences in the pattern of the pathological changes in scrapie, kuru, experimental kuru and subacute presenile polioencephalopathy, in Whitty CWM, Hughes JT, MacCallum FO (eds): *Virus Diseases and the Nervous System*. Oxford, Blackwell Scientific Publications, 1969, pp 107–120.

40. Chou SM, Martin JD: Kuru-plaques in a case of Cruetzfeldt-Jakob disease. *Acta Neuropathol* 17:150–155, 1971.

41. Hirano AH, Ghatak NR, Johnson AB, et al: Argentophilic plaques in Creutzfeldt-Jakob disease. *Arch Neurol* 26:530–542, 1972.

42. Gajdusek DC: Slow virus infection and activation of latent infections in aging. *Adv. Gerontol Res* 4:201–218, 1972.

43. Duffy P, Wolf J, Collins G, et al: Person-to-person transmission of Creutzfeldt-Jakob disease. *N Engl J Med* 299:692–693, 1974.

44. Gajdusek DC, Gibbs CJ Jr, Collins G, Traub RD: Survival of Creutzfeldt-Jakob disease virus in formol-fixed brain tissue. *N Engl J Med* 294:533, 1976.

45. Gajdusek DC, Gibbs CJ Jr, Earle K, et al: Transmission of subacute

spongiform encephalopathy to the chimpanzee and squirrel monkey from a patient with papulosis atrophicans maligna of Köhlmeier-Degos, in Subirana, A, Espadaler JM, Burrows EH (eds): *Proc X Int Cong Neurol, Excerpta Medica Int,* Congress Series No. 319. Amsterdam, 1973, pp 390–392.

46.   Matthews WB: Epidemiology of Creutzfeldt-Jakob disease in England and Wales. *J Neurol Neurosurg Psychiatry* 380:210–213, 1975.

47.   Bernoulli C, Siegfried J, Baumgartner G, Regli F, Rabinowicz T, Gajdusek DC, Gibbs CJ Jr: Danger of accidental person-to-person transmission of Creutzfeldt-Jakob disease by surgery. *Lancet* 1:478–479, 1977.

48.   Gajdusek DC, Gibbs CJ Jr, Asher DM, Brown P, Diwan A, Hoffman P, Nemo G, Rohwer R, White L: Precautions in medical care of and in handling materials from patients with transmissible virus dementia (Creutzfeldt-Jakob disease). *N Engl J Med,* in press.

14

Diagnostic Virology

CHARLES V. SANDERS, JR.

In 1949, Enders et al. (1) introduced methods for propagating poliovirus in cells of nonneural origin. Since then, almost 200 viruses that infect man have been isolated by *in vitro* culture techniques, and the spectrum of diseases recognized as being of viral origin has expanded. For example, since 1965, several neurological diseases of man have been shown to be caused by persistent unconventional viruses (see Chapter 13). More than 20 papovaviruses have been isolated, one of which, JCV, has been associated with progressive multifocal leukoencephalopathy, a rare demyelinating neurological disease found in some patients suffering from chronic leukemia, Hodgkin's disease, lymphosarcoma, or carcinomatosis (see Chapter 11). The propagation of viruses in cell culture has also made possible the quantitation of specific serum antibody, an equally important aspect of diagnostic virology.

The same clinical syndrome may be produced by many different viruses; conversely, a given virus may produce diverse clinical syndromes (Table 14.1) (2). The clinician should remember that mycobacteria, chlamydia, mycoplasma, fungi, and leptospira (all potentially treatable) may produce many of the syndromes (especially aseptic meningitis and encephalitis) listed in Table 14.1. Thus, establishing a diagnosis on clinical grounds alone is difficult, and the need for a diagnostic virology laboratory is apparent (3,4).

INDICATIONS FOR VIRAL STUDIES

The time and cost entailed in isolating and identifying viruses demand that clinicians use the diagnostic virology laboratory selectively. Confirming the diagnosis of a typical case of measles, mumps, or chickenpox or determining the cause of an upper respiratory tract infection in an individual patient has little practical value. What guidelines, then, should the physician use?

Virological studies may be justifiable in conditions in which:

1. The established diagnosis will directly affect the management of the patient.

Examples: For laboratory-documented rubella in the first trimester of pregnancy, a decision to terminate pregnancy may be considered. For a pregnant woman (at term) with active genital herpes simplex infection, cesarean section may be recommended. For either infection, relatively rapid and specific diagnosis is possible.

261

TABLE 14.1.

Clinical Presentation	Virus
Respiratory diseases	
Upper respiratory tract illness	Coronaviruses*; rhinoviruses*
	Parainfluenza*
	Respiratory syncytial*
	Influenza A, B
	Adenoviruses, echoviruses, coxsackie A 21, reoviruses
Exudative tonsillopharyngitis	Epstein-Barr virus
	Adenoviruses
Acute lymphonodular pharyngitis	Coxsackie A 10
Pharyngoconjunctival fever	Adenoviruses
Herpangina, stomatitis, pharyngitis, or all three	Coxsackie A 1–6,8,10
	Herpes simplex 1
Bronchiolitis	Parainfluenza*; influenza A
	Respiratory syncytial*
	Adenoviruses
Laryngotracheobronchitis (croup)	Parainfluenza*; influenza A, B†
	Respiratory syncytial
	Rhinoviruses
	Adenoviruses
Pneumonia	Respiratory syncytial*
	Rhinoviruses
	Adenoviruses

Specimens to Be Submitted to the Virus Laboratory

Specimen

Column legend:

1. Nasopharyngeal Washings
2. Nasal Swab
3. Throat Washing, Swab
4. Paired Sera
5. Nasopharyngeal Aspiration
6. Stool
7. Serum (Heterophil Agglutinins)
8. Swab of Oral Lesions
9. Lung Tissue
10. Vesicular Fluid & Scrapings (Cutaneous Lesions)
11. Fresh Urine
12. Acute-Phase Heparinized Blood
13. Eye Washings, Corneal or Conjunctival Scrapings
14. Saliva
15. Acute-Phase Clotted Blood
16. Liver & Spleen
17. Vaginal Swab
18. Cerebrospinal Fluid
19. Brain & Spinal Cord
20. Brain Tissue
21. CSF & Serum Hemagglutination Inhibition Antibodies
22. Cardiac Tissue
23. Tissue
24. Pleural Fluid
25. Synovial Fluid
26. Serum for Hepatitis B Surface Antigen (HBsAg)
27. Liver Tissue

1	2	3	4	5	6	7	8	9	10	11	12	13	14	15	16	17	18	19	20	21	22	23	24	25	26	27
+	+	+	+																							
	+	+																								
+	+	+	+																							
	+	+																								
	+	+		+																						
		+				+																				
	+	+		+																						
	+	+		+																						
	+	+		+																						
	+	+		+			+																			
	+	+					+																			
	+	+																								
	+	+	+	+																						
	+	+		+																						
	+	+																								
	+	+	+	+																						
+	+	+	+																							
	+	+		+																						
I	I	I	I				I,b,p																			
+	+	+	+																							
	+	+		+																						

TABLE 14.1.

Clinical Presentation	Virus
	Influenza; parainfluenza; rubeola
	Varicella
	Cytomegalovirus
	Herpes simplex
Influenza syndrome	Influenza A, B
	Influenza C
Ophthalmic diseases	
Keratitis	Herpes simplex type 1; herpes zoster
Epidemic keratoconjunctivitis	Adenovirus 8,* 19
Conjunctivitis	Newcastle disease virus; vaccinia
Exanthematous diseases	
Vesicular: Vesicular stomatitis with exanthem (hand, foot, and mouth disease)	Coxsackie A 5, 10, 16*
Vesiculopustular eruption	Varicella-zoster; herpes simplex 1, 2; vaccinia
	Variola
Maculopapular eruption	Rubella; rubeola
	Echovirus 4, 6, 9, 16; coxsackie A 9, 16, 23
Hemorrhagic fever	Dengue 1–4
Nonspecific febrile illness with exanthem	Echoviruses; coxsackie A, B
Genitourinary tract disease	
Vulvovaginitis	Herpes simplex 1, 2*
	Coxsackie B
Acute hemorrhagic cystitis	Adenovirus 11
Central nervous system disease	
Paralytic disease	Poliovirus 1,* 2,* 3*; coxsackie A 7, 9; echovirus 4, 6, 9
	Arborviruses
Aseptic meningitis	Mumps*
	Coxsackie A, B*; echoviruses*
	Western equine encephalitis
	Eastern equine encephalitis

								Specimen																	
2	3	4	5	6	7	8	9	10	11	12	13	14	15	16	17	18	19	20	21	22	23	24	25	26	27
	+	+						b,p																	
		+						b,p	+																
	+	+						b,p			+	+													
	+	+						b,p	+																
	+	+						b,p																	
	+	+																							
		+										+													
	+	+		+								+													
		+										+													
	+	+		+				+																	
		+						+																	
		+						+		+		+	+ p												
	+	+																							
	+	+		+																					
		+												+											
	+	+		+																					
		+						+								+									
	+	+		+				+								+									
		+							+																
	+	+		+													+								
		+																p							
		+								+		+				+									
	+	+		+												+									
		+																							
		+																							

TABLE 14.1.

Clinical Presentation	Virus
	Venezuelan equine encephalitis
	St. Louis encephalitis; California encephalitis
	Poliovirus 1, 2, 3
	Herpes simplex 1
	Herpes simplex 2
	Varicella-zoster
	Lymphocytic choriomeningitis
	Epstein-Barr‡
Meningoencephalitis	Rubeola*
	Mumps
	Western equine encephalitis*; Eastern equine encephalitis*
	Venezuelan equine encephalitis
	St. Louis encephalitis*
	California encephalitis*
	Cytomegalovirus
	Herpes simplex 1
	Herpes simplex 2
	Herpes-zoster
	Lymphocytic choriomeningitis
	Rabies
	Epstein-Barr
Guillain-Barré syndrome§	Coxsackie A; echoviruses
	Cytomegalovirus
	Epstein-Barr
Chronic CNS infections	
Subacute sclerosing panencephalitis (Dawson encephalitis)	Rubeola; rubella
Progressive multifocal leukoencephalopathy	Papovaviruses
Cardiovascular disease	
Myocarditis and pericarditis	Coxsackie B*; echoviruses
	Mumps

										Specimen																
1	2	3	4	5	6	7	8	9	10	11	12	13	14	15	16	17	18	19	20	21	22	23	24	25	26	27
		+	+											+			+									
			+																							
		+	+		+												+									
			+						+								+									
			+						+		+						+									
			+						+								+									
			+												+		+									
			+			+																				
																			b,p							
			+							+		+					+									
			+															p								
		+	+												+		+	p								
			+															p								
			+															p								
		+	+							+	+								b,p							
			+																b,p							
			+									+							b,p							
			+						+									p								
			+											+				p								
			+										+					p								
			+			+																				
		+	+		+												+									
		+	+							+	+						+									
			+			+																				
																			b,p	+						
																			b,p							
		+	+		+																p					
			+							+		+									p					

TABLE 14.1.

Clinical Presentation	Virus
Miscellaneous disease	
Cytomegalic inclusion disease	Cytomegalovirus
Parotitis	Mumps
Orchitis and epididymitis	Mumps*
	Coxsackie B
	Lymphocytic choriomeningitis
Infantile diarrhea	Rotaviruses
Epidemic myalgia (Bornholm disease, pleurodynia)	Coxsackie B 1–6
Nonspecific febrile illness	Polioviruses 1, 2, 3; coxsackie A, B; echoviruses
	Venezuelan equine encephalitis
Postperfusion syndrome	Cytomegalovirus*
	Epstein-Barr
Arthritis	Varicella-zoster
	Rubella
	Arboviruses
	Variola
Hepatitis (adult)	Hepatitis A*
	Hepatitis B*
	Cytomegalovirus
	Epstein-Barr
	Herpes simplex 1, 2
Neonatal	Rubella
	Herpes simplex 1, 2
	Coxsackie B

This table is an extensive modification of data given in ref. 12.
Key: +, from a living patient; I, in infants only; A, adults, especially females; b, biopsy; p, postmortem.
* Common cause of this condition.
† During epidemics, influenza virus may be the chief cause of severe croup.
‡ Infectious mononucleosis.
§ Guillain-Barré syndrome has followed immunization with measles vaccine and swine influenza vaccine (see Chapter 5).
|| By means of immune electron microscopy, viral particles have been demonstrated in stool specimens of patients having hepatitis A virus infections.

										Specimen															
2	3	4	5	6	7	8	9	10	11	12	13	14	15	16	17	18	19	20	21	22	23	24	25	26	27
	+	+							+	+											p				
	+								+			+													
	+								+			+													
	+	+		+																					
	+												+												
	+			+																					
	+	+		+																		+			
	+	+		+																					
	+	+											+												
	+	+							+	+															
	+				+																				
	+							+															+		
	A	A																				A			
	+																					+			
	+							+		+			+												
				‖																					
	+																							+	
	+	+							+	+															b,p
	+				+																				
	+																								b,p
	+	+							+																b,p
	+							+																	b,p
	+	+		+																					b,p

2. A correct diagnosis is vital to the community's health.

   Examples:

   a. A severely ill African visitor has a vesiculopustular eruption—is it smallpox or chickenpox?

   b. A patient has paralytic disease—is it due to poliovirus? An outbreak of poliomyelitis would lead to an intensification of immunization efforts.

   c. Influenza A is suspected—antigenic drifts may be determined so that new vaccines can be prepared.

   d. St. Louis encephalitis occurs in an urban area—control measures, such as aerial spraying, should begin.

   e. To protect community health, medical personnel in contact with an acutely ill patient must themselves be protected, for example, in infections with Lassa or Marburg viruses (see Chapter 8).

3. Knowing whether a disease that is "going around" in the community is of viral rather than bacterial origin may lead to restriction in the use of antibiotics, thus sparing patients the unnecessary expense and potential adverse side effects.

   Example: Antimicrobial therapy will neither benefit a patient with typical influenza nor prevent the development of secondary bacterial pneumonia, a frequent complication of influenza.

4. Knowing the cause of a disease is not only satisfying but also reduces unnecessary hospitalization, testing, and treatment.

   Example: Patients with viral aseptic meningitis will not usually require prolonged hospitalization.

5. Recent studies with adenine arabinoside (AraA) and interferon (see Chapter 16) suggest that the age of antiviral therapy is approaching. Rapid laboratory diagnosis and sensitivity testing will then be required as in antibacterial chemotherapy to assist in the selection of the proper antiviral drug.

   Example: At present, some viral diseases are being treated or prevented with antiviral agents, 5-iodo-2′ deoxyuridine (IDU), α-adamantanamine (amantadine), and N-methylisatin for prevention of influenza and β-thiosemicarbazone (methisazone) for prevention of smallpox.

6. Identification of a specific viral agent as the cause of clinical disease may be of prognostic value.

   Example: A child has encephalitis—is it due to mumps (low mortality) or eastern equine encephalitis virus (high mortality)?

7. The causative agent of a currently prevalent clinical syndrome is being sought.

Example: An outbreak of aseptic meningitis occurs—what agent is responsible?

8. Physician education is an imperative.

As stated by Douglas (5), "It seems certain that new manifestations of viral disease in man remain to be discovered, and this can only be accomplished by more frequent application of viral diagnostic procedures to clinical problems."

CLINICAL SPECIMENS

Many large medical centers employ a clinical virologist who is trained and has access to facilities for performance of viral diagnostic studies. If such a capability is not immediately available, a state health laboratory can usually supply this service. The physician wishing to send specimens for viral diagnosis should consult the laboratory to determine the type of specimens required and how to collect and transport them.*

Clearly, meaningful information as to isolation of viruses and pertinent serologic data cannot be derived from the laboratory examination if specimens are submitted unaccompanied by adequate clinical information. A specimen labeled "viral studies" is as useless as a blood sample submitted to the chemistry laboratory for "chemical studies."

Selection

The choice of specimens will depend on the clinical syndrome to be diagnosed or the virus suspected of causing the syndrome or both. The viruses associated with various syndromes and the clinical specimens that should be collected and submitted to the virus laboratory are listed in Table 14.1.

Example. Throat washings or lung tissue (biopsy or autopsy) are the specimens most likely to yield influenza virus during active disease (6).

---

* The Center for Disease Control has available a useful booklet, *Collection, Handling and Shipment of Microbial Specimens,* edited by Robert H. Huffaker and identified as DHEW Publication No. CDC 74-82630.

On the other hand, in patients with aseptic meningitis, culture of the cerebrospinal fluid may be negative, yet a virus may be recovered from other body sites (e.g., enterovirus from throat washings or stool, or mumps virus from urine or saliva.)

Collection

Specimens to be submitted for viral isolation should be collected as early as possible in the course of illness because, in most instances, the greatest amounts of virus are present during the prodromal stage and the first few days of illness. For example, influenza virus is rarely recovered from a throat swab or washings taken three or four days after the onset of illness (6). Herrmann and Herrmann (7) recovered influenza virus from throat specimens in 84% of samples taken one to four days after onset of illness but in only 13% of samples taken five to seven days after onset. On the other hand, after infection, some viruses may be shed for several weeks (enterovirus in the stool, adenovirus in the throat or stool) or even for months (cytomegalovirus in the urine or rubella virus in throat secretions of infants with congenital rubella syndrome) (8). One should remember that viruses, unlike most bacteria, are not viable when separated from host tissue. Some viruses are heat- and acid-labile (rhinoviruses), some are damaged by freezing-thawing (respiratory syncytial virus, CMV), while others are fairly stable (enteroviruses may be recovered from stool specimens maintained at room temperature for several days). The type of specimen and a brief description of the manner of collection are listed on the next page.

Storage

As a general rule, specimens should be held at 4°C if they are to be inoculated into cell culture within 24 hours of collection. If a specimen must be kept for a longer period before inoculation, it should be frozen at −70°C rather than at −20°C (9). Addition of protein (such as gelatin) to the carrier medium stabilizes relatively fragile viruses. Specimens that contain acid-labile viruses such as rhinoviruses and

COLLECTION OF SPECIMENS FOR VIRUS ISOLATION
AND OF BLOOD FOR SEROLOGIC TESTS

Note: All specimens should be collected *aseptically.* Cell culture medium will support growth of bacteria, which will destroy cell monolayers and thus preclude isolation of virus.

Blood*: Aseptically collect at least 20 ml of blood (10 ml of serum). Early separation of serum from clot prevents hemolysis, which interferes with some tests. Whole blood is often used for viral isolation; therefore, no preservative should be added. For some cell-associated viruses, heparinized or citrated blood may be inoculated into cell culture or animals or both, but using clotted blood and separating the serum is usually better. Do not freeze whole blood because hemolyzed blood will be toxic for cell cultures.

Nasopharyngeal washings: With the patient in the sitting position (head tilted back slightly) instill 4–5 ml of carrier medium† (without antibiotics) into each nostril. After 5 to 10 seconds, the patient brings his head forward and expels the fluid from his nostrils and mouth into a sterile screw-capped wide-mouth jar.

Nasal swab: Place a sterile dry cotton or Dacron swab into the patient's nostril and leave it in place for a few seconds, thus allowing for maximal absorption of nasal secretions. Repeat the procedure in the opposite nostril. Then dip the swabs into a vial containing 5 ml of carrier medium.

Nasopharyngeal aspiration: Introduce a no. 8 plastic disposable premature-infant feeding tube (attached to a 10-ml syringe) through the patient's nose into the nasopharynx. Secretions are aspirated and placed into 2–4 ml of carrier medium.

Throat washings: Have the patient gargle with 10 ml of saline solution or carrier medium (the latter is preferable) and expectorate into a sterile screw-capped wide-mouth jar. There is greater likelihood of isolating virus from a throat washing than from a throat swab.

Throat swabs: Rub a sterile dry cotton or Dacron swab across the patient's oropharynx and tonsils, and then dip the swab into a vial containing 5 ml of carrier medium.

Stool specimen or rectal swabs: Collect 5–10 gm of stool in a clean carton. This procedure is preferable to using rectal swabs because only small amounts of virus may be present. In using a rectal swab, pass the sterile swab, moistened with carrier medium, into the anus so that the cotton tip is no longer visible. Then withdraw and place the swab into 5 ml of carrier medium.

Body fluids (CSF, pericardial, pleural): Collect 3–5 ml of fluid in a sterile screw-capped glass tube.

Urine: Collect 10–30 ml of urine (midstream or catheter specimen) in a sterile container and store at 4°C until it can be examined.

Vesicle fluid or skin scrapings: Aspirate contents of the vesicle into a tuberculin syringe containing about 0.5 ml of carrier medium. Label syringe, cap it, and send it to the laboratory packed in ice. Alternatively, if no fluid can be aspirated, clean the vesicles with 70% alcohol, open them, and absorb exudate from several vesicles on two cotton swabs and place swabs in 1–2 ml of carrier medium.

Autopsy (or biopsy) specimens: Collect as soon as possible postmortem in a sterile tube or petri dish. Clinical syndrome will determine choice of tissue (e.g., brain, liver, lung).

Eye swabs: Swab conjunctiva with a sterile cotton swab and place the swab in a screw-capped tube containing 5 ml of carrier medium.

---

* The same blood may be used as an acute phase serum for serological tests.
† The carrier medium obtained from the virus laboratory usually consists of a balanced, buffered salt solution with added protein, such as gelatin, to stabilize the virus and antibiotics (penicillin and streptomycin or gentamicin) to suppress bacterial growth. *Note: A patient allergic to penicillin should not gargle with the carrier medium.*

arboviruses should be stored in dry ice and in sealed ampules to prevent acidification due to CO_2.

Transport

Specimens that are to be in transit to the laboratory for 24 hours or longer should be placed in tightly sealed containers securely packed in dry ice in an insulated (e.g., polyfoam) container.*

VIRAL INFECTIONS: CLINICAL AND EPIDEMIOLOGIC CLUES

A clinical and epidemiologic history is helpful to the physician in establishing presumptive diagnoses. This information in turn will aid the laboratory personnel in selecting the most likely host system for virus isolation and indicating appropriate serologic studies. The patient's age and place of residence are important in considering the most likely viral infections. Most patients with aseptic meningitis due to the enteroviruses are less than 20 years of age. Interpretation of clinical signs will narrow the diagnostic possibilities. For instance, data on the location, extent, and character of a vesiculopustular rash will aid in differentiating herpes simplex virus from varicella-zoster virus or smallpox (variola) virus infections. The patient's past history of infections is important because viral infections such as mumps or measles usually confer lifelong specific immunity. An epidemiologic history is also helpful. Encephalitis caused by St. Louis encephalitis virus has a seasonal peak of incidence in late summer and early autumn. Travel history is essential in the diagnosis of smallpox, dengue, and yellow fever because of their limited geographical distribution. A history of recent immunizations is helpful in excluding a diagnosis such as paralytic poliomyelitis or smallpox. Drug history may also provide valuable clues. For example, disseminated infections with CMV or VZV occur with some frequency in patients receiving immunosuppressive drugs.

---

* The most current information for shipment of diagnostic specimens is provided by the Biohazards Control Officer, CDC, 1600 Clifton Road, Atlanta, Ga. 30333; Phone: (404) 632-3311, ext. 3883.

LABORATORY DIAGNOSIS

Rapid, accurate laboratory diagnosis of viral infection depends on the compilation of information from several sources:

1. Clinical and epidemiologic history.
2. Direct microscopic examination of clinical specimens.
3. Direct isolation of a virus from appropriate clinical specimens.
There may be considerable variation in the time required for the laboratory to provide the physician with significant diagnostic information. The isolation of HSV from a clinical sample may take three days, whereas the isolation of the CMV may take two weeks. (As a general rule, the longer the time interval between submitting the specimen and receiving the results, the less likely a virus will be isolated from the specimen.)
4. Examination of paired (acute and convalescent) samples of serum demonstrating a significant (fourfold or greater) change in antibody titer to a given virus during the course of illness.

Virus Isolation and Interpretation of Results

Recovery of a virus from a patient with a particular clinical syndrome does not necessarily establish the diagnosis. On the other hand, many viral infections, especially in children, may be completely asymptomatic. The viral diagnostic laboratory may provide the specific diagnosis of a viral disease, for example, the recovery of HSV in disseminated neonatal disease or the isolation of CMV from the blood of a patient with fever of undetermined origin who has received a renal transplant.

In attempting to interpret the significance of viruses isolated from a patient the physician should consider:

1. The clinical syndrome and the particular virus isolated. The recovery of influenza virus from throat washings obtained from patients with an influenzal syndrome strongly suggests that the disease was caused by the agent. On the other hand, recovery of HSV from throat washings of a patient with pharyngitis does not necessarily indicate that

HSV caused the symptoms. It is possible that latent HSV may have been reactivated by some unrelated acute febrile illness and has been recovered coincidentally.

2. The source of isolated virus. The isolation of coxsackievirus from the cerebrospinal fluid of a child with meningitis has greater diagnostic significance than isolation of the same virus from feces. These and other enteroviruses may be shed continuously for several weeks after asymptomatic infection. However, in patients with aseptic meningitis, negative cultures of the cerebrospinal fluid for bacteria, the isolation of an enterovirus from the stool, along with demonstration of a four-fold rise in serum antibody titer directed against the recovered virus, provide sufficient evidence for establishing a diagnosis (10,11). Likewise, isolation of HSV from the brain of a patient with encephalitis establishes that diagnosis.

Serologic Tests

The serologic diagnosis of viral infection is based on the principle that almost all viruses stimulate the production of specific antibodies. It is inexpensive, but it necessitates a delay in diagnosis. In general, serologic studies are best performed on adequately spaced paired sera to demonstrate a change, usually a rise in antibody titer against the specific antigen of the suspected causative agent. The acute phase serum should be obtained when the patient is first seen and a convalescent serum specimen obtained 14 to 21 days later. If the acute phase serum sample is drawn too late, the antibody titer may have already peaked, and one may therefore fail to demonstrate a significant difference in titer between the acute- and convalescent-phase sera. Antibody levels determined on a single serum sample are usually not helpful because one cannot distinguish between recent and past infections. An important exception is rubella virus, antibody to which, discovered in a single serum sample, indicates immunity to rubella and is an especially important finding in pregnancy. Also, the finding of hepatitis B surface antigen (HBsAg) in a single serum specimen from a patient with a clinical syndrome of "viral" hepatitis would strongly suggest hepatitis B viral infection (see Chapter 4). The rapid laboratory diagnosis of viral infections is receiving increasing attention (12–16).

TABLE 14.2. *Procedures for Viral Serodiagnosis**

Antigen-Dependent	
Complement fixation	Immunoperoxidase-labeled antibody
Hemagglutination-inhibition	Indirect hemagglutination
Hemadsorption-inhibition	Platelet aggregation
Hemaggregation-inhibition	Radioimmunoassay
Immunodiffusion	Enzyme-linked immunosorbent (ELISA)
Immunoelectrophoresis	Immunofluorescence
Counterimmunoelectrophoresis	
Activity-Dependent	
Immune adherence	Neutralization (CPE, metabolic inhibition)
Specific lymphocyte stimulation (e.g., EBV)	

* Modified from ref. 14.

Procedures used for viral serodiagnosis are listed in Table 14.2. Viral diagnostic serology involves two choices:

1. Which virus should the laboratory be instructed to search for?
2. Which serologic method or methods should be used to examine paired serum samples?

For both, the clinical history is helpful in making a choice most likely to yield significant data. A request for "viral antibody screen" is too nonspecific, because testing is economically feasible for only a limited number of viruses that may be under greatest suspicion. For example, in patients with aseptic meningitis, such a request, without clinical and epidemiologic information, would require screening for more than 65 serotypes of enterovirus, an extravagant and meaningless task. The second choice (the serologic test) involves several considerations for both the clinician and the laboratory. Serum complement fixation and hemagglutination inhibition tests provide rapid answers and are inexpensive in comparison with neutralization tests that may require cell cultures and a week or more to provide an answer. On the other hand, the neutralization test is generally the most specific and is applicable to most viruses, whereas HI tests are applicable only to those viruses that agglutinate erythrocytes. In general, CF antibodies are specific and antibody titers fall rapidly (within months), whereas HI or neutralizing antibodies often persist unchanged for many months or even years.

277

GUIDELINES FOR THE CLINICIAN

The virology laboratory is more likely to provide meaningful data to physicians if these rules are observed:

1. For virus isolation, collect specimens from appropriate sources as early as possible in the course of the illness.
2. Freeze the specimen immediately below −70°C if it cannot be transported to the laboratory within 24 hours after collection.
3. Obtain acute and convalescent serum samples (paired sera).
4. The following represents the absolute minimum clinical information that must accompany each specimen submitted.
   a. Name, age, and sex of patient
   b. Summary of pertinent history, including date of onset of symptoms, physical findings, and pertinent laboratory data
   c. Virus disease suspected
   d. Type of material submitted and date and hour collected
   e. Number of similar cases in the family or neighborhood
   f. Viral vaccine(s) given to patient and date(s) of administration
   g. Extent of any exposure to animals or insect vectors

COMMENT AND CONCLUSIONS

For diagnostic virology to be rewarding, the physician and the virus laboratory must collaborate closely. If the laboratory fails to receive properly chosen, collected, and transported specimens, along with pertinent clinical information, much hard work by both physician and laboratory may be fruitless. With the advent of effective viral chemotherapeutic agents, the need for rapid virological diagnosis and antiviral sensitivity testing will increase. Diagnostic virology facilities, like conventional microbiology laboratories, are now included in many major health centers.

Many procedures in the virology laboratory are not only expensive but also cumbersome and time-consuming. Thus physicians often do not receive a quick response after they have submitted the appropriate specimens. However. continued study with several of the tests, that is, counterimmunoelectropuoresis, immunoperoxidase-labeled antibody, enzyme-linked immunosorbent assay (ELISA), and immunofluorescence, show that they hold great promise for rapid diagnosis.

DISCUSSION

Question: What special precautions do the speakers take in their laboratories in their respective disciplines?

Dr. Gajdusek: As required by NIH (National Institutes of Health), all work with human and animal agents is done under high containment. That is certainly not the case for most of the brain tissue handled outside our laboratories. In other words, we're getting material that has been borne up and down the corridors of American hospitals by biochemists who have no autoclaves in their departments, and from small pieces of such tissue we're transmitting the disease. I would say that this circumstance exists in every hospital in the United States having an academic department of biochemistry or physiology. The fear, then, is not that these agents may be widely disseminated around the country from viral laboratories that use sterile techniques and isolation of their animals but from butcher shops and autopsy rooms and even surgical suites throughout the United States.

Dr. Khoury: With regard to containment facilities, we work within the guidelines proposed by NIH. I'd like to answer this question in two parts. First, I have the advantage of working in a containment building. With regard to SV40 virus, I think it's survived the test of time, and it's sufficient to work with it if proper precautions are taken—no mouth pipetting, laminar flow hoods, essentially P2 type facilities. For some of the newer human agents that we don't know enough about yet, I think we'd be more circumspect and use, for example, P3 facilities, which in addition require negative pressure rooms, gloves, and so forth. That's not to say that the newer agents are more dangerous. I personally believe that they may be less dangerous. We certainly have antibody levels against them that we don't have against SV40. I think, however, that nobody can argue against stringent precautions when an agent isn't really well studied and hasn't been tested for some time.

The second part of my answer relates to some of the experiments we're going to be involved in using the recombinant DNA molecules and, of course, with them the requirements are much more stringent and certainly will be adhered to. Just a word of caution about that type of experiment: I think that the concern many people have is firmly founded, given some of the descriptions that Dr. Gajdusek has presented on infectivity of viroids and some of the recent evidence from our laboratory and other laboratories at NIH that indicate the viral DNA from papovaviruses themselves may be infectious when injected intraperitoneally or subcutaneously. The DNA clearly is infectious, and genes can be expressed. These really will require the ultimate precautions to be used when recombining new DNA molecules that might be infectious.

Dr. Robb: I've had three experiences that have made me exceedingly cautious. First, several years ago, when I was at NIH, we mouth-pipetted without cotton-plugged pipettes. I took a mouthful of SV40 of fairly high titer and converted in two weeks. I took it home and apparently transmitted it through my saliva to my wife, and she converted as well. We haven't checked our children yet, but that experience scared me.

The second incident happened when I had two students come into my laboratory, which is now under P2½ containment, with laminar-flow hoods, negative pressure, disinfection of everything, changing shoes and gloves, and so forth. There, also, the students became positive within three weeks. They had broken the techniques we had developed, although they were doing everything we told them to do. Fortunately, SV40 appears to be safe in the long run. It is very antigenic and unless you're immunosuppressed, you won't get into trouble.

The third experience is ongoing in that I'm creating mutants, and the "Andromeda strain" is always in the back of my mind. Fortunately, it turns out that any ts (temperature-sensitive) mutant we isolate in the laboratory has probably occurred in nature. The danger is that we grow them up to very high titers and work with them in the laboratory. Here again, I chose a virus that grows only in specialized mouse strains and that does not infect humans, and we do have antibodies that cross-react to it. So, you try to select a virus system that is very safe and also handle it with good containment. Do not take the need for precautions lightly. My laboratory will not allow any students to come in who have not had sound and thorough training in safe virology techniques somewhere else.

REFERENCES

1. Enders JF, Weller TH, Robbins FC: Cultivation of the Lansing strain poliomyelitis virus in cultures of various human embryonic tissues. *Science* 109:85–87, 1949.
2. Fenner FJ, White DO: Laboratory diagnosis of viral disease, in Fenner FJ, White DO (eds): *Medical Virology*, ed 2. New York, Academic Press, 1976, pp 255–285.
3. Herrmann EC, Herrmann JA: Survey of viral diagnostic laboratories in medical centers. *J Infect Dis* 133:359–362, 1976.
4. Horstmann DM: Clinical virology. *Am J Med* 38:739–750, 1965.
5. Douglas RG: Laboratory diagnosis of viral and mycoplasmal infections, in Knight V (ed): *Viral and Mycoplasmal Infections of the Respiratory Tract*. Philadelphia, Lea and Febiger, 1974, pp 11–21.
6. Schaeffer M: Diagnosis of viral and rickettsial diseases, in Miller SE (ed): *Textbook of Clinical Pathology*, ed 8. Baltimore, Williams and Wilkins, 1971, pp 329–366.
7. Herrmann EC, Herrmann JA: Laboratory diagnosis of viral disease, in Drew WL (ed): *Viral Infections—A Clinical Approach*. Philadelphia, FA Davis, 1976, pp 23–45.
8. Lennette EH: General principles underlying laboratory diagnosis of viral and rickettsial infection, in Lennette EH, Schmidt NJ (eds): *Diagnostic Procedures for Viral and Rickettsial Infection*, ed 4. New York, American Public Health Association, 1969, pp 1–65.
9. Walker WE, Martins RR, Karrels PA, et al: Rapid clinical diagnosis of common viruses by specific cytopathic changes in unstained tissue culture roller tubes. *Am J Clin Pathol* 56:384–393, 1971.
10. Balfour HH Jr: Practical virology: Indications and yield. *Minn Med* 58:421–422, 1975.
11. Berlin BS: Some facets of clinical virology. *Comprehensive Therapy* 1:64–73, 1975.
12. Gardner PS, McQuillin J (eds): *Rapid Virus Diagnosis—Application of Immunofluorescence*. London, Butterworths, 1974.
13. Benjamin DR: Use of immunoperoxidase for rapid viral diagnosis, in Schlessinger D (ed): *Microbiology—1975*. Washington DC, American Society for Microbiology, 1975, pp 89–96.
14. Kapikian AZ, Dienstag JL, Purcell RH: Immune electron microscopy as a method for detection, identification, and characterization of agents not cultivable in an in-vitro system, in Rose NR, Friedman H (eds): *Manual of Clinical Immunology*. Washington DC, American Society for Microbiology, 1976, pp 467–480.
15. Voller A, Bidwell D, Bartlett A: Microplate enzyme immunoassays for the immunodiagnosis of virus infections, in Rose NR, Friedman H (eds): *Manual of Clinical Immunology*. Washington DC, American Society for Microbiology, 1976, pp 506–512.
16. McCracken AW, Newman JT: The current status of the laboratory diagnosis of viral diseases of man. *CRC Crit Rev Clin Lab Sci* 5 (3):331–363, 1975.

15

Immunization Against Viral Infections

DOROTHY M. HORSTMANN

Providing basic immunization has long been the province of the pediatrician. The American Academy of Pediatrics accepted the responsibility for supplying authoritative information on the subject and has done so in the form of its so-called Red Book (Report of the Committee on Infectious Diseases), first issued in 1938. The guidelines provided have served as the basis of immunization practices relating to children for nearly 40 years. If all children were adequately immunized and all vaccine-induced immunity were long-lasting, physicians caring for adults would have little need to be concerned, except under special circumstances. But the matter is not so simple: Certain segments of the population escape immunization in childhood, vaccine immunity is not always long-lasting nor is it as protective as natural immunity, and changes in the epidemiology of certain diseases occur as a result of immunization practices. These facts make it necessary for all physicians to be alert to current needs if optimum control is to be maintained. The degree of success achieved so far is different for various diseases. The following review is a progress report as to where we now stand with regard to prevention of viral infections of major significance in the United States. Because influenza and hepatitis are reviewed elsewhere (see Chapters 4 and 5), they will not be discussed here, but I shall make a few comments concerning rubella, a subject covered more completely in Chapter 6.

VACCINE PROPERTIES

Before discussing immunization against specific viral infections, the types of vaccines currently in use, their advantages, and their disadvantages should be mentioned. Table 15.1 lists some of the properties that are essential for an ideal vaccine. The properties range from the immunogenic capacity of vaccines in terms of humoral and cellular immune responses and absence of hazards associated with their use to acceptance by the public and cost effectiveness. Three types of vaccines meet these qualifications in some degree: killed-virus preparations; live, attenuated-virus vaccines; and nonreplicating subunit vaccines. Examples of each are listed in Table 15.2.

Killed-Virus Vaccines

In the preparation of killed-virus vaccines, chemical treatment is used in a way calculated to destroy infectivity of the agent without destroy-

TABLE 15.1. *Properties of an Ideal Viral Vaccine*

1. Induces a full complement of protective humoral antibodies
2. Stimulates local IgA antibody production in sites where entry of the virus occurs
3. Induces cell-mediated immunity
4. Results in immunity equal to that acquired naturally in terms of degree and duration of the above immune responses
5. Is not associated with side reactions of significance
6. If a live-virus preparation, does not back-mutate and acquire unacceptable virulence
7. Does not result in enhanced virus persistence and is not oncogenic
8. Can be given in a form that is simple logistically, is acceptable to the public, and fits in with the practices of private practitioners and of public health physicians
9. Can be justified on a cost-benefit analysis in terms of cost of vaccine and immunization versus cost of disease in terms of morbidity, mortality, and emotional impact.

ing antigenicity. Examples are rabies vaccine, in which the virus is inactivated by B-propiolactone, and influenza and poliovirus vaccines, which are treated with formalin. Killed antigens provide stimulation of the humoral immune system, with resultant production of specific circulating antibodies. In general, the antibody titers induced are lower than those that follow immunization with live-virus vaccines; although use of adjuvants can increase the levels, the associated local reactions that sometimes occur (e.g., nodules, cysts) are generally unacceptable complications. An advantage of killed-virus vaccines is their safety

TABLE 15.2. *Viral Vaccines—1976*

Killed	Live	Subunit (Nonreplicating)
Rabies	Smallpox	Influenza (split)
Influenza	Yellow fever	Hepatitis B*
Polio	Polio	Adenovirus*
	Measles	Rabies*
	Mumps	
	Rubella	
	Adenovirus—types 4 and 7†	
	Respiratory syncytial*	
	Influenza*	

* Experimental
† For military recruits

286

when properly prepared and used. Disadvantages are the generally less solid immunity induced, the failure of some vaccines to stimulate a full spectrum of antibody responses, and the tendency of antibody titers to decline to low levels that are not protective. Other problems are the absence of or the relatively poor response induced in terms of local IgA antibody on mucous surfaces of the nasopharynx and alimentary tract, with resultant lack of resistance to infection at the first line of defense. In addition, certain killed-virus vaccines such as those against measles and respiratory syncytial virus induce an altered immune state that favors accentuation of disease on exposure and infection with wild virus (1).

Live-Virus Vaccines

Among the live-virus vaccines, smallpox vaccine has been so successful that the need for it has disappeared, except in Ethiopia. Continued vigilance will be needed to ensure this control. The other viral vaccines noted in Table 15.2 represent strains of various viral agents that have been attenuated by successive passages in animals or tissue culture, with resultant loss of virulence but with retention of the ability to replicate in humans and to induce immunity. Genetic stability and lack of communicability are desirable features that characterize some, but not all, of the vaccines listed.

Live-virus vaccines have the advantage of simulating natural infection. Inoculation or ingestion of an attenuated strain is followed by virus multiplication locally and at distant sites, thus providing a far larger antigenic mass that can be administered in 0.5 or 1.0 ml of a killed-virus vaccine. An important feature of the immunity induced is that it depends on both circulating antibody and local resistance associated with the presence of secretory IgA antibody on the mucous surfaces which are the site of primary implantation of virus in many infections. More information is needed to establish whether attenuated strains induce satisfactory cell-mediated immunity.

A potential hazard is inherent in live-virus vaccines: Multiplication of the attenuated strain may result in progeny that acquire more virulent properties, which render them capable of causing disease. Another concern is virus persistence. Will the ultimate consequences of infection be different with attenuated than with naturally occurring virus strains? Enhancement of oncogenicity by attenuated strains of certain agents of the herpes group such as EBV and HSV-2 is also a possible problem, although at present the hazard is a theoretical one. Evaluation of the issues of persistence and oncogenicity is difficult because only long-term observation of humans can adequately answer these questions.

Subunit Vaccines

Rapidly advancing knowledge of the structure of viruses has provided information that can be used to develop subviral or subunit immunogens for use in prophylaxis (2). Such vaccines are obtained by disrupting the viruses and purifying the antigens responsible for stimulating antibodies. The currently licensed influenza split-virus vaccine consists of disrupted virions but is not a true subunit preparation. In the case of hepatitis B vaccine, which is still in the experimental stage, advantage is taken of the surface antigen that is already present in high titer as a subunit of the virus in the blood of carriers; purified HBs antigen thus serves as a vaccine. Experimental subunit vaccines have also been prepared against adenovirus type 7 and against rabies. The rabies vaccine has given encouraging results in preliminary trials in man.

Highly purified preparations of the viral antigens responsible for inducing immunity have the advantages of safety and lack of toxicity. Their ultimate effectiveness is as yet unknown, and as with other new developments, unforeseen problems may appear. Nevertheless, the potential for effective, safe subunit vaccines warrants further effort to develop them.

IMMUNIZATION AGAINST SPECIFIC DISEASES

Poliomyelitis

Paralytic poliomyelitis has been virtually eliminated from large geographical areas as a result of the introduction of vaccines and their extensive use during the past 20 years. Yet epidemics continue to occur in various parts of the world, particularly in developing countries where economic and other problems have delayed institution of adequate control measures. In the United States, the disease has been reduced remarkably: The total annual number of paralytic cases has fallen from an average of about 21,000 cases in the five years before introduction of vaccine in 1955, to 7 cases in 1973, 7 in 1974, and 8 in 1975 (3). This is indeed a remarkable record of success—a success that is far greater than was anticipated even by the most optimistic.

The inactivated poliomyelitis vaccine (IPV), the first to be licensed, is given in three or four injections during a period of six to eight months. It was used extensively in the United States until the mid-1960s with considerable resultant decline in incidence of the disease, although epidemics continued to occur. In several small European countries, the vaccine is still used with great success. Oral poliomyelitis vaccine (OPV), licensed in 1961–1962, subsequently was recommended as the vaccine of choice in the United States, and in recent years, it has supplanted the killed-virus preparation, which is not currently available in this country.

At present, trivalent oral poliovirus vaccine is recommended for routine use in infants and young children; for adults, it is recommended only under special circumstances, such as for travel to areas where poliomyelitis is highly endemic. Advantages of the live-virus vaccine include ease of administration and speed of the immune response. It induces both circulating antibody and local IgA antibody, which provides resistance of the intestinal tract to reinfection and thus decreases the circulation of virus in the community. Inactivated poliovirus vaccine also induces circulating antibody that protects against CNS invasion, but because it stimulates little or no local IgA antibody production, it is less effective in preventing multiplication of the virus in the intestinal tract.

Despite the extraordinary success of the oral vaccine, some problems remain. One is the failure to reach more than 65% of the nation's preschool children, especially those who live in inner-city and other poverty areas. The causes of this poor record are multiple. Young parents, to whom the devastations resulting from epidemics of poliomyelitis in prevaccine days are unknown, tend to be apathetic about having their children protected against the disease. Also, health workers often have difficulty in finding and identifying the target preshool-aged populations, as well as in educating parents. Decrease in federal and state support for vaccination programs, such as the termination of the National Vaccine Assistance Act, has also played a role. The overall result has been a shift in the epidemiologic behavior of poliomyelitis in the United States, with a concentration of cases due to wild virus occurring in the 0 to 4 year age group and the virtual absence of cases in the school-aged children 5 to 14 years—the group that formerly had the highest incidence of the disease.

Other problems with oral vaccines have to do with the vaccine viruses themselves, with inadequate host responses in certain parts of the world, largely the developing countries, and with the declining serologic immunity rates among children in certain large cities in the United States and elsewhere.

Oral Vaccine Virus Mutation to Virulence. The immunogenic effectiveness of oral vaccine depends on multiplication of the attenuated virus in the intestinal tract. Because no poliovirus strain is completely stable, progeny of the vaccine strains regularly undergo a certain degree of mutation; in rare instances, such mutations have resulted in increased virulence and in vaccine-associated cases in recipients and their close contacts, often their parents (3). The actual risks are difficult to assess accurately because the denominators are unknown, but for the period 1965–1973, when trivalent OPV was used almost exclusively, the estimated rate of vaccine-associated cases in recipients was 1 in 11.5 million; in contacts it was 1 case for each 5 million doses given (4). These data are based on specific criteria defining vaccine-associated cases as those occurring in recipients within 7 to 30 days and in contacts within 7 to 60 days after administration of the vaccine. In addition, laboratory tests are used to characterize the virus isolated from such cases as either "vaccinelike" or "wild." The tests cannot definitely establish the origin of the strains, however, nor do they tell anything about the neurovirulence of the virus recovered. Nevertheless, when taken together with the epidemiologic features involved, they can aid in determining the probable source of an infection. The experience in the United States with probable vaccine-associated cases is paralleled in certain other parts of the world, as indicated by a recent WHO collaborative study (5).

The risks associated with oral vaccines are exceedingly small, and in view of the benefits provided, they are considered acceptable in most parts of the world. Nevertheless, in the United States the question has recently been raised as to whether the inactivated poliovirus vaccine should be used again. Experience in this country indicates how difficult it is to reach a high percentage of children with oral vaccine; the likelihood is that the record would be far worse with a killed-virus vaccine that requires multiple injections over a period of months. In developing countries the logistics of administering such a vaccine make its effective use virtually impossible. Although the rare vaccine-associated cases that occur in the United States would be eliminated by a return to killed-virus vaccine, the cost might well be a far larger number of paralytic cases among the unvaccinated or inadequately vaccinated young children. Wild polioviruses are still extant, as was illustrated in 1972, when an epidemic of poliomyelitis occurred in a small Christian Science school in Connecticut (6). Almost all the children were unvaccinated, and 11 paralytic cases occurred among the 129 pupils. This experience highlights the potential danger of concentrations of susceptibles in a world in which the circulation of wild polioviruses has been greatly reduced but clearly not eliminated.

Certain small, highly developed countries like Sweden have excellent records of prevention of poliomyelitis through use of killed-virus vaccine only. Such records have been achieved because of the homogeneous nature and small size of the populations and the efficiency of the health services that regularly succeed in reaching virtually 100% of the children with three to four injections of vaccine (7). In a large, diverse country like the United States, an equivalent success is not likely.

However, there is one strong indication for the use of IPV in the United States, and that is in persons with hypogammaglobulinemia. Live-virus vaccine is contraindicated in such individuals because of their increased risk of paralytic disease: Nearly 10% of the vaccine-associated cases reported between 1961 and 1971 were in persons with abnormal immunoglobulins, an incidence estimated to be 10,000 times that for the normal population (8).

Problems with Oral Vaccine in the Tropics. A persistent difficulty with the use of live poliovirus vaccine among populations living in tropical and semitropical areas has been the greatly reduced seroconversion rates among susceptible children (9,10). Thus, success rates have been as low as 50%, in contrast with the almost 100% response that is characteristic of programs in the United States and other Western countries. Viral interference from infection with another enterovirus at the time of vaccine ingestion (a frequent occurrence) does play some role in this problem but apparently not an important one. Rather, the presence of an inhibitory substance in the oropharynx of the children has been found to prevent viral multiplication (10). In a field study conducted in Africa under the auspices of the World Health Organization, this inhibitor was found to prevent significant multiplication of the vaccine virus. The effect can be partially neutralized by antibodies present in horse serum prepared against human gamma globulin, and the oral administration of this serum at the same time as the vaccine enhances virus multiplication and greatly improves seroconversion rates. In another approach, John et al. (11) have recently reported greatly improved seroconversion rates by using 10 times the usual dose and giving each type separately rather than in repeated doses of trivalent vaccines. In view of these several findings, a modification of current poliovirus immunization practices in developing countries would seem to require consideration.

Decline in Immunity Rates. Serologic surveys have been used to advantage in monitoring immunity to poliomyelitis. In 1963, shortly after the introduction of oral vaccine and its enthusiastic reception, Melnick et al. found that a high percentage of children in Houston, Texas, were immune to all three types of poliovirus, as evidenced by significant neutralizing antibody titers. Five years later, however, the picture had changed, and a considerable increase was found in the percentage of schoolchildren who apparently had not been vaccinated and who were susceptible to one or more types (12). That antibody patterns differ in different areas is shown by the results of serologic surveys during 1965 and 1970 in New Haven, Connecticut, where a high immunity rate has been maintained and during 1970 in Syracuse, New York, where susceptibility rates resembled those in Texas (Fig. 15.1) (13). Others have noted similar trends (14,15). Gold et al. (14) recently found that a surprising number of Cleveland preschool children had antibody titers of $< 1 : 10$ against one or more poliovirus types, despite a history of having received oral vaccine.

The absence of epidemics in the face of these findings is attributable to several factors, both known and unknown. Decreased circulation of wild poliovirus strains after the widespread use of oral vaccine is no doubt of primary importance. The lack of aggregation of a sufficient number of susceptibles in a setting of exposure to wild virus and

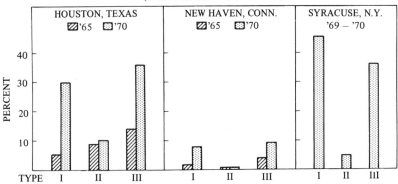

Figure 15.1. Absence of poliovirus neutralizing antibody in children five to nine years old (serum titers <1 : 5 or <1 : 8). (From ref. 13.)

of close personal contact for efficient spread is another. A third more speculative reason is the probability that resistance of the intestinal tract is maintained in vaccine recipients whose circulating antibody titers have fallen below the 1 : 5 or 1 : 10 level, which is the lowest serum dilution commonly used in neutralization tests. This interpretation is supported by evidence indicating that resistance to reinfection has been observed in natural immunes who lack detectable antibody, and this phenomenon increases with increasing age. Nevertheless, a low level or absence of antibodies in a significant segment of the childhood population is a matter of concern and deserves to be followed closely. Consideration might well be given to putting greater emphasis on revaccination of all children entering school using trivalent vaccine as recommended by the American Academy of Pediatrics.

Measles

Immunization against measles became a reality in 1963, 10 years after the virus was first isolated. The vaccine has now been given to about 80 million children, and its use has had a notable impact on the incidence of the disease in the United States (Fig. 15.2). The number of reported cases fell from 458,000 in 1964 to a low of 22,000 in 1968.

MEASLES – US, 1963 – 1976

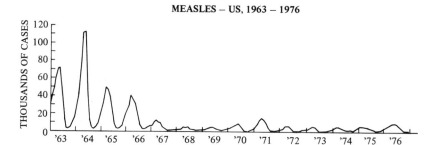

Figure 15.2. Measles cases reported to United States Center for Disease Control, Atlanta, Georgia, 1963–1976.

This reduction was accompanied by a concomitant decline in measles encephalitis. Yet in 1975, approximately 26,000 cases of measles were reported, and already more than 40,000 have been notified in 1976. As with vaccination against poliomyelitis, failure to reach poverty-area children is in part responsible for the continued occurrence of cases and localized epidemics.

The current recommendation for primary immunization against measles is that the live attenuated virus vaccine be given at 15 months of age (16). Two vaccine preparations are generally available, Schwarz and Attenuvax; both represent "further attenuated measles vaccine," produced by additional passage of the original Edmonston strain in chick fibroblast tissue culture at lowered temperature. The vaccines induce inapparent infection in recipients; little or no virus shedding occurs, and the infection is not communicable. Minor side effects occur in 10% to 20% of children. These consist of fever and a faint rash 5 to 10 days postimmunization. No neurological complications have been documented, but a variety of CNS syndromes after immunization have been reported; the available evidence suggests that they are coincidental and not pathogenically associated with the vaccine. A few cases of SSPE have also been reported in children who have been immunized, but the risk is lower after measles—1 per million in contrast to 5 per million after the disease.

Successful immunization with live attenuated measles virus vaccine is followed by the appearance of protective antibody responses in about 95% of recipients. In several studies, persistence of immunity for at least the 6 to 14 years that the vaccine recipients have been followed has been documented (16,17). The standard serologic test used is

294

hemagglutination inhibition. Krugman (16) has shown that patterns of decline in HI antibody titers in children who received the original Edmonston B strain are similar to those that follow natural measles; however, titers are about two- to four-fold lower in children who received further attenuated vaccines that have been widely used since 1968. The overall results of long-term observation have been interpreted as probably indicating life-long immunity after injection of live measles virus vaccine. Despite such optimism, a number of unforeseen problems have arisen concerning the immunity induced by both the killed-virus and live-virus measles vaccines.

Atypical Measles after Killed-Virus Vaccine. The occurrence of atypical measles in children who had received killed-virus vaccine was observed as early as 1965, when this product was in use either alone or with live-virus vaccine. It was subsequently withdrawn when it became apparent that when exposed to wild measles virus, a number of children immunized with the killed preparation developed severe, atypical illnesses suggestive of delayed hypersensitivity reactions (1). These consisted of high fever, an unusual rash (sometimes purpuric) that began on the extremities and spread cephalad, and severe pneumonia. Furthermore, when children who had received the killed vaccine were subsequently given the live attenuated virus as was recommended beginning in 1968, they frequently developed local erythema, tenderness, swelling, and a vesicular or hemorrhagic lesion at the site of injection. Although the mechanism of these reactions is not understood, part of the explanation for the failure of inactivated vaccine to protect against infection has recently been suggested by Norrby et al. (18), who have demonstrated differences in the antigenic structure of live and killed measles virus vaccines. Thus, both wild measles and attenuated live virus induce antibodies to two different surface-envelope components of the virus—the hemagglutinin and the hemolysin. In contrast, killed virus (whether inactivated by formalin or by treatment with TWEEN 80 and ether) induces antibody only to the hemagglutinin and not to the hemolysin.

Measles after Live-Virus Vaccine. As more experience has accumulated during the 13 years since introduction of measles vaccine, the fact that a history of vaccination does not necessarily mean immunity has become increasingly apparent. Thus, cases of measles have been reported in children under several circumstances. First, about 5% of children fail to get a successful "take" on immunization and develop classical measles when exposed to the natural disease. A substantial proportion of these cases has been in children who were given vaccine at 9 to 12 months of age, when persistent maternal antibody was sufficient to prevent a primary antibody response and successful immunization. A second group of children, which includes both those vaccinated before 1 year of age and some immunized at 12 months and older, has also been found to develop measles on exposure to wild virus, with antibody responses that indicate that they *had* responded to primary immunization but had *lost* protective immunity. Linneman et al. (19) conducted a serologic study of an epidemic in a highly immunized population in which the overall attack rate was 20.9% for those who had not been immunized and 10.6% for those who had received vaccine, either live or killed. The protective efficacy was 49% in those immunized when less than 1 year of age and 83% in children who had received vaccine at 12 months or older. The clinical features in the vaccinated and unvaccinated groups were not markedly different, but the convalescent HI antibody titers of those who had been vaccinated were far higher than in the unimmunized children. Among those who had received only live-virus vaccine, the geometric mean titer in convalescent sera was 1 : 1372, whereas for the unvaccinated it was 1 : 190. These results suggests that the vaccine recipients had responded to the original immunization, but their immunity had declined to unprotective levels; the high titers after reinfection thus represented anamnestic responses in previously sensitized persons. This interpretation was confirmed on examination of the sera of some of the children, which revealed a characteristic secondary-type antibody response, that is, IgG only, and no detectable IgM antibody (20).

Similar observations have been made by others. Schluederberg et al. (21) have reported a secondary-type antibody response in an 8-year-old child who had measles encephalitis and had received Edmonston B vaccine plus gamma globulin at 10 months of age. Cherry and his colleagues (22) have described six cases of measles with secondary-type antibody responses in children who had been immunized with live virus only at 10 months, 11 months, 1 year, and 5 years of age. The complexities of the situation are illustrated by the observation that these children, unlike the ones described by Linneman et al., experienced atypical measles—severe disease comparable to that reported earlier in children who had been given killed-virus vaccine. In any event, the findings indicate that the capacity for anamnestic antibody response is not necessarily protective against clinical measles, thus belying a cherished assumption.

With these reports and other similar ones at hand, the question arises as to whether more and more cases of measles will be seen in vaccine recipients as the time since the first widespread use of the vaccine increases. Recent outbreaks, such as that occurring in 1976 in Connecticut (A. Piccirillo and D. Smith, personal communication) and involving children who had been immunized up to 11 years earlier with Schwarz vaccine, suggest that vaccine recipients might not have prolonged immunity. Weiner et al. (23) reported that 35% of 54 cases in vaccinees reported in an epidemic in New York State, were in adolescents who had received live measles virus vaccine without immune globulin after one year of age. The trend toward such outbreaks raises questions about the durability of vaccine-induced immunity, particularly that following use of the further-attenuated vaccines, which are less immunogenic than the original Edmonston B strain.

Problems in the Diagnosis of Measles. As a result of widespread immunization, the age incidence of measles in certain areas has shifted, as illustrated by the recent experience in California (Fig. 15.3) (24). This shift has involved two groups of persons: adults who have escaped immunization and have been protected from natural exposure as a result of decrease in circulation of the virus, and vaccine recipients whose immunity has declined after 10 to 12 years to unprotective levels and who later developed atypical measles on exposure to wild virus. Both groups have presented diagnostic problems. During an epidemic in the San Francisco Bay area, Rand et al. (24) observed four cases in unimmunized adults who presented with unusual clinical pictures: Two of the patients, both on immunosuppressive drugs, died with giant-cell pneumonia, and a third died with postinfectious encephalitis. During 1974 and 1975, atypical measles accounted for 56 of 177 laboratory-confirmed cases in California (25). The ages of the patients ranged from 2 to 24 years, and 47 of the 56 had in the past received either killed-virus vaccine alone or killed followed by live-virus vaccine within three months; one had received only live-virus vaccine. The mean interval between vaccination and onset of disease was 10½ years, and the longest was 12½ years. A correct clinical diagnosis of

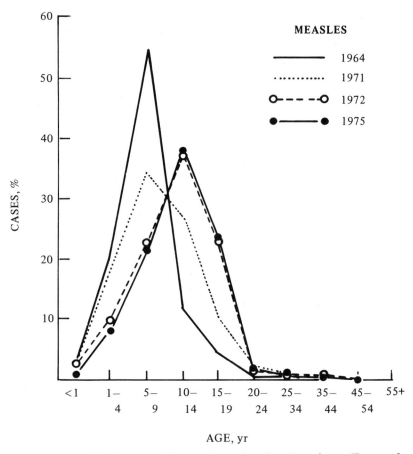

Figure 15.3.   Shift in age incidence of measles, San Francisco. (From ref. 24.)

measles was made in only 29% of patients; other diagnoses considered included Rocky Mountain spotted fever, influenza, and staphylococcal pneumonia. Many young physicians today have never seen a case of measles. The importance of their being aware of the disease and of its potential for unusual behavior in young adults is becoming more and more apparent.

Should Revaccination Against Measles be Recommended? Reinfection of vaccine recipients with wild virus had been demonstrated to occur and has been viewed as probably desirable as long as it does not result in disease. With the decline in wild-virus circulation, however, opportunities for enhancing immunity by this means have diminished correspondingly. An alternative mechanism is to revaccinate with live-virus vaccine. Bass and his colleagues (26) recently undertook an investigation of the effectiveness of this procedure. Children immunized with live virus up to 8 years previously were compared in terms of antibody status and response to revaccination with those who had experienced natural measles up to 20 years before. The results indicated that antibody titers were considerably lower in vaccine recipients than in natural immunes. If an HI titer of $< 1 : 2$ was taken to indicate lack of immunity, 8.6% of the vaccine recipients fell into this category two years postvaccination, and the figure rose to 27.5% in children vaccinated more than eight years previously. Most of these persons did have neutralizing antibodies at a titer of 1 : 2, and only 15, or 5%, of the 318 vaccinated children lacked *all* detectable serologic evidence of immunity. In contrast, HI and neutralizing antibodies were present and in considerably higher titer in all 49 of the natural immunes, even 15 to 20 years after infection. Not surprisingly, those vaccinated proved to be far more susceptible to reinfection with the vaccine strain than were those who had experienced natural infection: None of the natural immunes showed any change in antibody level, regardless of prebooster titer, whereas 29% of the vaccine recipients had $\geq$ fourfold rises. However, rises in neutralizing antibody were maintained in only a quarter of the children who were tested six months later, raising some question as to the extent that revaccination augments immunity.

The results of Bass et al. indicate a considerably greater decline and loss of antibody among vaccinees than has been reported by Krugman (16) and by Weibel et al. (17). Thus, Bass found 27.5% to be HI negative 8 years postvaccination when sera were tested at a 1 : 2 dilution, whereas Krugman found only 9% negative in similar tests of vaccine recipients 12 years after vaccination. Apparently vaccine-induced immunity may not be as long-lasting as anticipated. With the increasing numbers of clinical cases of the disease being reported in persons immunized 10 years and more previously, long-term clinical and serologic studies are indicated to determine the possible need for revaccination and the optimum time after primary immunization that revaccination might be given.

Mumps

Immunization against mumps has been relatively slow in gaining widespread acceptance, largely because the disease is far less important than polio and measles as a cause of significant morbidity and mortality. The infection may be inapparent, but it commonly presents as a mild disease of young school-aged children, 90% of cases occurring in those less than 14 years of age. As with other "childhood" infections, adults commonly suffer a more severe clinical course and a higher rate of complications, including encephalitis, aseptic meningitis, orchitis, and damage to the VIIIth cranial nerve with resultant deafness. Myocarditis and pancreatitis also occur, and evidence suggests that involvement of the pancreas may be associated with early-onset diabetes. Therefore, good reason seems to exist for preventing mumps if an effective vaccine is available.

The history of immunization against mumps has been reviewed by Deinhardt and Shramek (27). It began in the 1940s, when Enders induced protection against experimental infection in monkeys by inoculation with an attenuated strain of the virus. Shortly afterward, Henle et al. prepared a formalin-inactivated vaccine that induced circulating antibodies and protected against the disease but did not prevent inapparent infection. Subsequent evidence indicated that the killed-virus vaccine failed to provide reliable long-term immunity, and attention was therefore turned to a live-virus vaccine. In 1967, the Jeryl-Lynn strain of mumps virus, which had been attenuated by Buynak and Hilleman by passage in eggs and in chick embryo tissue cultures, was licensed in the United States. The U.S. Immunization Survey conducted in 1974 indicated that 39% of children one to nine years of age had received the vaccine, a considerably lower figure than the approximately 65% to 70% ratio for polio, measles, and rubella vaccines. In the future, as mumps is more often included with measles and rubella in the multivalent vaccine given to infants, the percentage immunized will increase.

Experience with live attenuated virus mumps vaccine has shown that it is well tolerated; adverse reactions are rare and consist mainly of mild parotitis. In a few instances, CNS abnormalities, including encephalitis, have been reported as occurring between two days and three weeks after vaccination. The small numbers of cases and the spread in terms of time of onset suggest that if CNS complications are associated with immunization, the event must be rare.

Mumps vaccine induces inapparent infection and antibody rises in 95% of susceptible recipients. So far, vaccinees have been protected against the disease on subsequent contact exposure to persons with natural mumps (17). The potential effectiveness of widespread use of the vaccine was demonstrated in Massachusetts, the only state to implement a full-scale immunization program (Fig. 15.4). A remarkable fall in incidence was observed as an increasing proportion of the childhood population was immunized during the period 1968 to 1972. By 1975, only 308 cases were reported. Although the national decline in incidence of mumps is not as striking as in Massachusetts, it is nevertheless impressive (28) (Fig. 15.5). The decline has been accompanied by a similar reduction in mumps encephalitis, from an average

Figure 15.4. Reported cases of mumps, Massachusetts, 1968–1973; cumulative doses of mumps vaccine administered. (From Center for Disease Control—Mumps Surveillance, January 1972–June 1974, issued October 1974.)

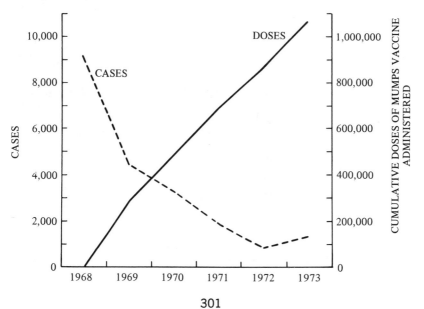

IMMUNIZATION AGAINST VIRAL INFECTIONS
MUMPS — US, 1960 — 1974

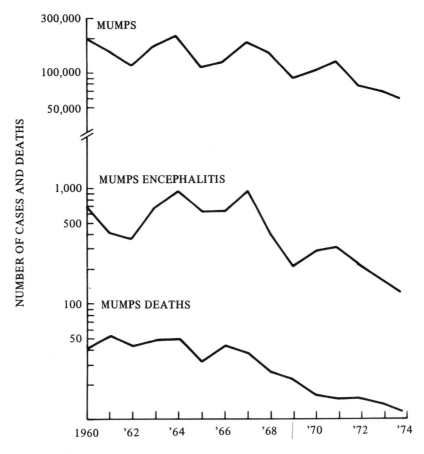

Figure 15.5. Mumps, mumps encephalitis, and deaths due to mumps. (Modified from Center for Disease Control—Mumps Surveillance, January 1972–June 1974, issued October 1974.)

of 743 reported cases annually during the five years before licensure of the vaccine to 149 in 1975.

The present recommendations are that children be immunized at 15 months of age if a combined measles, mumps, rubella preparation is used. The vaccine is also recommended in the absence of a history of the disease for older children approaching puberty and for adults, particularly men. The duration of immunity is not known, but experience so far is encouraging in that no loss of protection has occurred by eight years after vaccination. However, continuing surveillance is necessary to be certain that immunization in childhood will prevent the disease and its complications in those exposed to the natural infection many years later as adults.

Rubella

Immunization against rubella, as noted by Schiff (see Chapter 6), has had a decided impact on the incidence of the disease in the United States. Epidemics have declined sharply, and fewer cases of the congenital rubella syndrome have been reported. The overall picture is encouraging, but problems have arisen that indicate that caution is necessary in predicting the elimination of rubella.

One problem has to do with the quantity—and quality—of the resistance induced by vaccines as compared with that following natural infection. We have been much concerned with these aspects in a long-term follow-up study of about 500 susceptible children who received the $HPV_{77}DE_5$ vaccine in a prelicensure field trial in 1968 (29). In brief, we found that those children who had brisk antibody responses (HI, neutralizing, CF, and precipitating) after vaccination, maintained their serologic immunity without significant decline over a five-year period. They constituted two thirds of the vaccine recipients. In contrast, the one third whose serologic responses were less vigorous, as indicated by HI titers in the 1 : 8 to 1 : 16 range two months after vaccination, showed a loss of detectable HI antibody in 25% during the ensuing three to five years. Because the neutralizing antibody (NT) is probably the protective one, neutralization tests were performed on sera of 25 vaccine recipients whose HI levels had declined to < 1 : 8. The results are shown in Table 15.3. One third of the children had low-level

TABLE 15.3. *Neutralizing Antibody Status of Rubella Vaccinees that Lost HI Antibody after 3 to 5 years*

HI Titer 2 mo Postvaccine	Number Neutralizing AB Positive/ Number Tested		
	After 2 mo	After 3 yr	After 5 yr
1:8	1/5	0/5	0/4
1:16	5/16	4/16	2/10
1:32	3/3	3/4	2/4

NT two months after vaccination, and 4 of these 18 remained positive five years later. The question arises as to how well-protected these children are and how well-protected they will be 10 or 15 years from now. Inherent in such a question is the concern that should reinfection occur during pregnancy, waning immunity might not be adequate to protect against viremia and fetal involvement.

A recent observation by Forsgren (personal communication) in Sweden has some bearing on this point. A woman who had both HI and CF but no detectable NT was inapparently reinfected with wild rubella virus during pregnancy; she gave birth to an infected infant who shed virus from the throat and had rubella IgM antibody. Clearly, low-level HI (1 : 32) with no NT did not protect in this instance against wild virus infection, viremia, and involvement of the fetus. The possible significance of this finding is illustrated by our observation that, overall, 10% of those with low-level HI antibodies lacked NT three years after vaccination. In contrast, among the natural immunes tested, all had NT even though their HI titers were low. This difference may well be part of the explanation of the far greater reinfection rate in vaccinated persons than in natural immunes.

Predictably, reinfection with wild rubella virus occurs most frequently in those with low antibody levels. Figure 15.6 illustrates the serologic responses of nine successfully vaccinated children who were reinfected. The geometric mean titers of the various antibodies—particularly neutralizing ones—after experience with wild virus reflect the more solid immunity that natural infection induces. How frequently such reinfection will occur in the future depends on how extensively wild rubella virus continues to circulate. That this circulation has been greatly reduced by the current immunization program in the United States is clear but that the virus will be eliminated is highly unlikely. Immunization surveys indicate that only 65% of children in the United States have received rubella vaccine. The unvaccinated group, plus those who have lost detectable immunity over the years, form a susceptible pool that is likely to keep the virus around for some time. Apparently, we are not yet in a position to be complacent about prevention of congenital rubella. Prospects should improve, however, when the RA27/3 vaccine of Plotkin is licensed, for the immunity it induces is more like that following natural infection in terms of breadth of antibody response, stimulation of local IgA in the oropharynx, persistence of serologic immunity, and resistance to challenge (30,31).

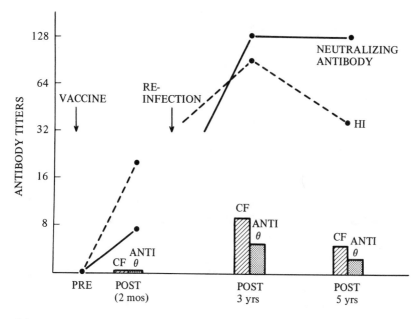

Figure 15.6.   Serologic responses of rubella vaccinees on natural reinfection with wild virus. GMT, geometric mean titers; HI, hemagglutination inhibiting; CF, complement fixing; anti θ, anti-theta antibodies.

Respiratory Infections

Morbidity from the common cold and other respiratory tract infections currently accounts for the greatest number of days lost from school or work and the greatest economic loss in our society. Controlling these infections by means of immunization has not been successful except in a few isolated instances. Thus, orally administered live adenovirus types 4 and 7 vaccines have been effective in greatly decreasing acute respiratory disease (ARD) among recruits in military camps. These viruses are of little importance in civilian life, accounting for only 3% or so of acute respiratory illnesses in young adults. The vaccines are therefore recommended only for military recruits.

Respiratory syncytial virus (RSV) is an important cause of severe and sometimes fatal bronchiolitis and pneumonia in young infants; the peak incidence is in those two to three months old. To prevent such

305

cases caused by RSV, immunization of children in the first month of life would be necessary. A killed-virus vaccine, the first to be developed, was found to induce antibodies in young infants but was ineffective in preventing infection and disease. Field trials were ended when children exposed to natural RSV infections some months after receiving the vaccine experienced a disease of greatly increased severity at an age when the infection is usually extremely mild. Subsequently, Chanock et al. (32) produced an attenuated strain of virus that will grow only at low temperatures, such as those in the human nasopharynx, and will not propagate at the higher temperature of the lungs. The attenuated strain is thus a temperature-sensitive mutant, designated ts-1. When given intranasally to adults the vaccine caused a localized inapparent infection and induced antibody production and resistance to infection with wild virus. The virus shed in the course of the infection remained stable and retained its ts characteristics. A different response has been encountered in susceptible infants given ts-1: They developed a mild inconsequential rhinitis, but in a few, the rhinitis was followed by otitis media, which made the vaccine unacceptable. In addition, the virus shed from the oropharynx by some of the children was found to have been changed genetically, since it was able to grow at temperatures up to 39°C. Further genetic alteration of the ts-1 mutant is necessary before an acceptable avirulent and stable RSV vaccine can be achieved. Work toward this end is in progress in Dr. Chanock's laboratory.

Common Cold Viruses. Among the most prominent pathogenic agents of minor upper respiratory tract infections are the rhinoviruses. They exist in more than 100 serotypes, and the immunity they induce is largely type-specific. This is a discouraging state of affairs indeed in terms of approaching control through immunization. A slight ray of hope is provided by recent observations by Cooney et al. (33) that some shared antigens exist among certain types, and a potent monovalent vaccine given to rabbits will induce heterotypic as well as homotypic antibodies. Also, epidemiologic evidence indicates that relatively few of the large number of serotypes are responsible for the bulk of infections at any one time. This suggests the possibility that a multivalent vaccine might be reasonably effective if the pattern is a repetitive one. Because parenteral injection does not induce local IgA antibody in the nasopharynx, such a vaccine would have to be given intranasally.

Clearly, we are a long way from successful immunization against rhino-virus infections.

The situation is no better with regard to the coronaviruses, which also contribute significantly to the incidence of common colds and other respiratory tract infections, particularly in the period from January to May each year. These viruses, with few exceptions, grow only in organ cultures and do not produce cytopathic effects. Their identification is largely by means of electron microscopy or immunofluorescence microscopy. This makes both virus isolation and serologic tests cumbersome. Information about the biologic behavior and other aspects of the coronaviruses is currently unfolding; however, it will be some time before enough is known to consider immunization against these agents.

Parainfluenza viruses, types 1, 2, and 3 cause primarily mild upper respiratory tract infections, but they are also responsible for croup in the first two years of life (para 1 and 2) and for bronchiolitis and pneumonia in infants six months of age and younger (para 3). Efforts to develop vaccines against these agents were begun as soon as their clinical importance was documented. Formalin-inactivated preparations given parenterally induced satisfactory antibody rises in suscepti-bles but unfortunately this was not correlated with protection against in-fection with wild virus (34). Resistance is apparently more closely associated with local secretory IgA antibody in the nasopharynx than with circulating antibody. Attempts to develop attenuated temperature-sensitive mutants similar to the RSV ts-1 vaccine strain have been un-dertaken, but the efforts are still in a preliminary stage.

Herpesviruses

This group of agents is responsible for a wide variety of human infec-tions; the viruses include varicella-zoster, cytomegalovirus, Epstein-Barr virus, and herpesvirus types 1 and 2 (see Chapters 9 and 10). The spectrum of host responses induced by these agents ranges from acute diseases, such as chickenpox and infectious mononucleosis, to inap-parent infections and latent infections that are periodically reactivated. In addition, at least one member of the group, EBV, shows a regular association with two human cancers, African Burkitt's lymphoma and nasopharyngeal carcinoma, and three members, HSV, CMV, and EBV, are able to initiate some type of cell transformation. Epidemiologic evidence of a possible relationship between HSV-2 and cervical cancer also exists, but the relationship is much weaker than that between EBV and African Burkitt's lymphoma.

Immunization against the array of clinical problems associated with the herpesviruses would seem to be highly desirable. Successful attenuated live-virus vaccines have been developed against herpesvirus infections of animals, such as Marek's disease of chickens, fowl laryngotracheitis, and infectious bovine tracheitis. Many problems are associated with translating experience with these vaccines in birds and animals to the control of infections in man in view of the incomplete state of knowledge with regard to the natural history and pathogenesis of the viruses in the human host (35). Furthermore, no satisfactory animal model exists in which to test human HSV vaccines because the behavior of this group in heterologous species differs considerably from their behavior in natural hosts. Also, in terms of possible live-virus preparations, such characteristics as persistence, latency, and reactivation may conceivably be enhanced in attenuated strains of the various agents.

A live attenuated VZV vaccine has recently been developed in Japan by serial passage of the agent in tissue culture (36). In a field trial in which it was given to contacts of cases of chickenpox, the vaccine was shown to induce antibodies and to protect against the disease (37). Despite this success, there is no great enthusiasm for such a vaccine (38) for several reasons: Chickenpox in normal children is usually a mild disease; severe cases and the rare deaths occur primarily in immunosuppressed hosts in whom the infection could be prevented by passive immunization with human immune serum globulin. Furthermore, it is not known what effect the vaccine might have on the occurrence of zoster, a result of late reactivation of the virus that does not develop until some years after the primary infection. No markers are available to determine the propensity of a given strain to cause zoster, and it would require several decades of observation to find out whether the vaccine would be associated with less—or more—severe zoster.

Despite this bleak picture, newer approaches, such as the use of purified subunit antigens, may prove successful in the development of vaccines against some diseases caused by the herpesviruses. An EBV vaccine to prevent infectious mononucleosis, an important disease in young adults, would have appeal; long-term complications would not be a significant cause for concern because reactivation of the virus is not associated with disease, and the possible oncogenic potential of the agent is limited by environmental and host factors that do not prevail in the United States.

IMPORTANCE OF SURVEILLANCE

Obviously, many more viral vaccines exist, actual and potential, that have not even been mentioned. But with regard to present and future vaccines, one point should be reemphasized, namely, the importance of continuous surveillance to determine the effectiveness of the various vaccines and the impact of their use on the epidemiology of the infections as well as on the diseases they cause (13). Such surveillance requires not only improved reporting by physicians of information on morbidity and mortality but also periodic serologic surveys to answer adequately pertinent questions concerning reinfection without clinical disease, the persistence of antibodies and of immunity in vaccine recipients, and the long-range results of vaccination programs, including their impact on circulation of the respective viral agents. With such information at hand, identification of potentially dangerous trends evidenced by increasing rates of susceptibility in given segments of the population should be possible. Such identification, in turn, would provide a sound basis for altering immunization practices (and perhaps immunizing agents) so that optimum vaccines can be more effectively delivered to those persons who need them most.

DISCUSSION

Question: Will it be desirable to reimmunize children with measles vaccine who were immunized at 12 months instead of 15 months?

Dr. Horstmann: Yes. The current recommendation is that children who were immunized before 13 months of age be considered for revaccination. This could be done routinely at the time of school entry. However, if epidemics involving previously vaccinated adolescents continue to occur, a more extensive program might have to be undertaken. We should perhaps be prepared for this.

Work by the author was supported by USPHS grant AI-11788 from the National Institute of Allergy and Infectious Diseases.

REFERENCES

1. Report of a Workshop: Disease accentuation after immunization with inactivated microbial vaccines. *J Infect Dis* 131:749–754, 1975.
2. Rubin BA, Tint H: The development and use of vaccines based on studies of virus substructures. *Prog Med Virol* 21:144–157, 1975.
3. Schonberger LB, McGowan JE Jr, Gregg MB: Vaccine-associated poliomyelitis in the United States, 1961–1972. *Am J Epidemiol* 104:202–211, 1976.
4. Center for Disease Control: Neurotropic Diseases Surveillance. Poliomyelitis. Annual Summary, 1973, issued February 1975.
5. Melnick JL: Enteroviruses, in Evans AS (ed): *Viral Infections of Humans.* New York, Plenum, 1976, p 196.
6. Center for Disease Control: Epidemiologic notes and reports. Followup on poliomyelitis—Connecticut. *Morbidity and Mortality Weekly Report* 21 (44):373–374 (Nov 4), 1972.
7. Böttiger M, Zetterberg B, Salenstedt CR: Seroimmunity to poliomyelitis in Sweden after the use of inactivated poliovirus vaccine for 10 years *Bull WHO* 46:141–149, 1972.
8. Wyatt HV: Poliomyelitis in hypogammaglobulinemics. *J Infect Dis* 128:802–806, 1973.
9. John TJ, Jayabal P: Oral poliovaccination of children in the tropics: 1. The poor seroconversion rates and the absence of viral interference. *Am J Epidemiol* 96:263–269, 1972.
10. Dömök I, Balayan MS, Fayinka OA, Skrtïc N, et al: Factors affecting the efficacy of live poliovirus vaccine in warm climates. Efficacy of type 1 Sabin vaccine administered together with antihuman gamma-globulin horse serum to breast-fed and artificially fed infants in Uganda. *Bull WHO* 51 (4):333–347, 1974.
11. John TJ, Devarajan LV, Balasubramanyan A: Immunization in India with trivalent and monovalent oral poliovirus vaccines of enhanced potency. *Bull WHO* 54:115–117, 1976.
12. Melnick JL, Burkhardt M, Taber LH, et al: Developing gap in immunity to poliomyelitis in an urban area. *JAMA* 209:1181–1185, 1969.
13. Horstmann DM: Need for monitoring vaccinated populations for immunity levels. *Prog Med Virol* 16:215–240, 1973.
14. Gold E, Fevrier A, Hatch MH, et al: Immune status of children 1–4 years of age as determined by history and antibody measurement. *N Engl J Med* 289:231–235, 1973.
15. Oberhofer TR, Brown GC, Monto AS: Seroimmunity to poliomyelitis in an American community. *Am J Epidemiol* 101:333–339, 1975.
16. Krugman S: Present status of measles and rubella immunization in the United States: A medical progress report. *J Pediatr* 90:1–12, 1977.
17. Weibel RE, Buynak EB, McLean AA, et al: Long-term follow-up for immunity after measles monovalent or combined live measles, mumps, and rubella virus vaccines. *Pediatrics* 56:380–397, 1975.
18. Norrby E, Enders-Ruckle G, terMeulen V: Differences in the appearance of antibodies to structural components of measles virus after immunization with inactivated and live virus. *J Infect Dis* 132:262–269, 1975.

19. Linneman CC Jr, Rotte TC, Schiff GM, et al: A seroepidemiologic study of a measles epidemic in a highly immunized population. *Am J Epidemiol* 95:238–246, 1972.

20. Linneman CC Jr, Hegg ME, Rotte TC, et al: Measles IgM response during reinfection of previously vaccinated children. *J Pediatr* 82:798–801, 1973.

21. Schluederberg A, Lamm SH, Landrigan PJ, et al: Measles immunity in children vaccinated before one year of age. *Am J Epidemiol* 97:402–409, 1973.

22. Cherry JD, Feigin RD, Lobes LA Jr, et al: Atypical measles in children previously immunized with attenuated measles virus vaccines. *Pediatrics* 50:712–717, 1972.

23. Weiner LB, Corwin RM, Nieburg PI, Feldman HA: A measles outbreak among adolescents. *J Pediatr* 90:17–20, 1977.

24. Rand KH, Emmons RW, Merigan TC: Measles in adults: An unforeseen consequence of immunization? *JAMA* 236:1028–1031, 1976.

25. Center for Disease Control: Current Trends. Atypical measles—California, 1974. *Morbidity and Mortality Weekly Report* 25 (31):245–246, (Aug 13), 1974.

26. Bass JW, Halstead SB, Fischer GW, et al: Booster vaccination with further live attenuated measles vaccine. *JAMA* 235:31–34, 1976.

27. Deinhardt F, Shramek GJ: Immunization against mumps. *Prog Med Virol* 11:126–153, 1969.

28. Center for Disease Control: Current Status of Mumps in the United States. *J Infect Dis* 132:106–109, 1975.

29. Horstmann DM: Controlling rubella: Problems and perspectives. *Ann Intern Med* 83:412–417, 1975.

30. Plotkin S, Farquhar JD, Ogra PL: Immunologic properties of RA 27/3 rubella virus vaccine: A comparison with strains presently licensed in the United States. *JAMA* 225:585–590, 1973.

31. Grillner L: Immunity to intranasal challenge with rubella virus two years after vaccination: Comparison of three vaccines. *J Infect Dis* 133:637–641, 1976.

32. Chanock RM, Kim HW, Brandt C, et al: Respiratory syncytial virus, in Evans AS (ed): *Viral Infections of Humans: Epidemiology and Control.* New York, Plenum, 1976, pp 365–382.

33. Cooney MK, Kerry GE, Tam R, et al: Cross relationships among 37 rhinoviruses demonstrated by virus neutralization with potent monotypic rabbit antisera. *Infect Immun* 7:335–340, 1973.

34. Chin J, Magoffin RL, Shearer LA, et al: Field evaluation of a respiratory syncytial virus vaccine and a trivalent parainfluenza virus vaccine in a pediatric population. *Am J Epidemiol* 89:449–463, 1969.

35. Parks WP, Rapp F: Prospects for herpes vaccination—safety and efficacy considerations. *Prog Med Virol* 21:188–206, 1975.

36. Takahashi M, Okuno Y, Otsuka T, et al: Development of a live attenuated varicella vaccine. *Biken J* 18:25, 1975.

37. Asano Y, Nakayama H, Yazaki T, et al: Protection against varicella in family contacts by immediate inoculation with live varicella vaccine. *Pediatrics* 59:3–7, 1977.

38. Brunell PA: Protection against varicella. *Pediatrics* 59:1–2, 1977.

16

Chemotherapy

GEORGE GEE JACKSON

THE DEVELOPMENTAL ARENA

Background and Progress

Although all chemotherapy contains an element of empiricism, the most successful application depends on knowing about the specific cause of an infection and understanding the pathogenesis and viral replicative processes. Successes in antibacterial chemotherapy have shown the potential for specific microbial inhibition and have created an expectancy about the development of successful antiviral chemotherapy. One limiting factor has been the inability to readily recognize specific viral infections by clinical and easy laboratory identification. Virology as a discipline, with the methods for determining specific strains of viruses, the pathogenesis of disease, and some of the molecular biology of viral replication, is of recent origin and still being developed. Methods for the use of tissue cultures to recognize viruses associated with certain diseases only became available at about the time that penicillin became a prime treatment for bacterial infections. Even now, diagnosis by virus isolation in tissue cultures has only limited availability and use. Thus, virology and the informational basis of antiviral chemotherapy are in their infancy.

Antiviral chemotherapy, although similar to antibacterial chemotherapy in principle, is biologically different. Bacteria usually multiply in an extracellular site, have two types of nucleic acid, divide by binary fission, and have a botanical cell envelope. These properties, lending an advantage to antibacterial chemotherapy, are not applicable in considering inhibition of viruses by drugs. Viruses have only a single type of nucleic acid, are dependent for replication on a viral-directed diversion of normal cellular metabolism and an integrated intracellular synthesis of various components at different production sites, and have no cell wall. Obviously, the biochemical discrimination of any antiviral agent must be more precise in specifically differentiating closely related viral and cellular processes than has been the rule with antibacterial or antiparasitic treatment. On the other hand, the additional steps of virus attachment, penetration, assembly, and release offer new sites for interruption by antiviral chemotherapy.

Because of the stringent requirements for an antiviral drug imposed by the intimate interrelation of the viral and cellular processes, one can adopt either an optimistic or pessimistic view with regard to the success, or lack thereof, in developing drugs for clinical use against viral diseases. Three compounds have been licensed for clinical use in the treatment of viral infections in man. They are a thiosemicarbazone, idoxuridine, and amantadine (Fig. 16.1). The first, methisazone (N-methyl-isatin β thiosemicarbazone [Marboran]) is used in the prevention of smallpox and the treatment of complicated vaccinia (1–3). It is already obsolete owing to the near eradication of smallpox and the change in vaccination policy. Idoxuridine (5-iodo-2′-deoxyuridine

Figure 16.1.   Three antiviral drugs licensed for clinical use.

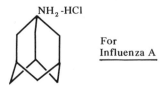

1−methylisatin−3−thiosemicarbazone
(methisazone, Marboran®)

5−iodo−2′−deoxyuridine (IDU)
(idoxuridine, Stoxil®)

1−adamantanamine hydrochloride
(amantadine, Symmetrel®)

[Stoxil]) is used as an ophthalmic solution at a concentration of 1 mg/ml (0.1%) in sterile distilled water and as an ophthalmic ointment at 0.5%. It has been effective in the topical treatment of ulcerative herpesvirus keratitis (4–6). Amantadine (1-adamantanamine hydrochloride [Symmetrel]) is used in the prophylaxis of influenza A (7). Numerous other compounds have demonstrable antiviral activity *in vitro* or in experimental models of viral infections, but their toxicity precludes clinical use, or they have not yet been fully studied.

The Role and Need for Drugs in the Control of Viral Infections

With the limited number of available drugs for a large variety of viral diseases, one must consider where antiviral drugs fit into the scheme of treatment and control of viral infections. The first opportunity is in prevention. Although immunoprophylaxis is preferred when available, antiviral drugs may be useful in the prevention of infections when the exposure of nonimmune persons to overt clinical cases of contagious infections is unavoidable. Although the use of vaccine, elimination of reservoirs of infection, and suppression of vectors have been successful methods for control of the more severe epidemic diseases, they are of little or no use for some of the milder viral diseases and those in which reinfection and latency occur. The prophylactic use of drugs for specific viral infections is logistically difficult, and noncompliance by well persons decreases drug effectiveness. However, prophylaxis may offer the best opportunity for the beneficial use of antiviral drugs. Time constraints on prophylaxis require quick and easy access to the drug (8), and the regimens must be safe when given for prolonged periods or in repeated courses. Progress in the use of prophylactic antiviral drugs will require new practices and procedures for making them more readily available with information about their safe use and safeguards against their abuse.

When the symptoms of viral infection become evident, chemotherapy is more readily sought and is the only specific measure with the potential to alter the natural course of the disease. By that time, however, the viral infection may have already induced severe and sometimes irreversible changes in infected cells. If so, a true therapeutic effect is difficult

317

to achieve. Containment, repair of viable cells, and the relief of acute symptoms may be the best that can be expected, even from a premier antiviral drug.

The needs for antiviral chemotherapy are great. Viral respiratory infections have high priority because of their prevalence and extensive morbidity. Viral infections of the central nervous system occur frequently and, if untreated, can be severe and damaging. Recrudescences of herpetic eruptions and virus-induced papillomas are prevalent viral diseases for which effective therapy would be beneficial. Systemic viral infections, such as hepatitis and, in some parts of the world, the hemorrhagic fevers and others, are diseases for which effective treatment measures are needed and unavailable (see Chapter 8). More remote, but perhaps of great potential importance, is effective antiviral chemotherapy for infections with some defective, persistent, and slow or unconventional viruses (see Chapter 13).

The Evaluation of Antiviral Therapy

In the perspective of the broad gauge of the needs and the variety of viral diseases for which treatment is needed, one should not expect that a single agent or even several drugs will have the spectrum of antiviral utility that is necessary and desired. Progress in antiviral chemotherapy, therefore, is likely to follow the course of step-by-step identification of specific needs, uses, and mechanisms of each new antiviral compound with potential for clinical use. A major problem in the development of such drugs is their clinical evaluation. Even the common viral diseases are seldom well enough understood to be predictable in terms of their pathogenesis, course, and duration of excretion of infectious virus. Treating infected persons in open field trials can produce strong opinions and few firm data about the usefulness of antiviral chemotherapy. Well-controlled clinical tests involving virologic, serologic, and clinical measurements are essential. Perhaps even before that, more clinical and virologic investigation is needed of the diseases themselves. Within the present social, political, and economic restraints on new-drug investigation, critical evaluations of antiviral chemotherapy will continue to be difficult and the documentation of new-drug effectiveness slow.

A further important aspect of the evaluation of antiviral chemotherapy is consideration of the host factors that participate in making a person susceptible to an infection and affect its course. Although most viral infections occur in otherwise normal persons, many involve young infants or persons who have some degree of impairment of immuno-competence owing to age, condition, or the administration of drugs. These circumstances are important in the evaluation and expectations from chemotherapy; in the prevention or treatment of recurrent endogenous latent infections, the immunologic status of the host may be paramount.

Mechanisms of Antiviral Drugs

The study of antiviral chemotherapy provides a double opportunity, one to use drugs to obtain new information about viral infections and the other to control them. Already, antiviral chemotherapy has made contributions that have expanded the classical concepts of how drugs prevent or cure infections and have introduced new ones. Some of these are given as categorical mechanisms in Table 16.1.

In the development of chemotherapy, a logical beginning is the use of chemicals that inactivate the infectivity of virus on direct contact. Methods used for lesions that are directly accessible are cryotherapy, chemical cauterization, irradiation with ultraviolet light, and photoinactivation after the application of heterocyclic dyes (9). The mechanism of drug action is through the cross-linkage of viral DNA or denaturation of capsid proteins. Difficulties include the need for repeated

TABLE 16.1. *Categorical Mechanisms of Action of Antiviral Drugs Used in Man*

Direct effect
Contact inactivation of viruses
Surface—envelope effect
Alteration of the virus-cell interaction
Chelation of trace metals, $Mn^{++}Mg^{++}$
Metabolic effect
Inhibition or misdirection of macromolecular
synthesis—DNA, RNA, protein, other
Host effects
Induction of resistance in host cells

applications, the production of transient inflammation, scar formation, inconsistent efficacy, and, in the case of photoinactivation, persistence of some functions of the viral genome, including the capacity to transform infected cells (9–11).

In the treatment of systemic infections with drugs that act primarily by viral inactivation, obtaining adequate concentrations at the right site for a sufficient period is difficult. Several compounds have been tried, but they have been ineffective. Some that inactivate influenza viruses on exposure but have failed in clinical trials are glyoxals (12), ABOB [N'N'-anhydrobis- (B-hydroxyethyl) biguanide hydrochloride] (13), substituted isoquinolines (14), and calcium elenolate (15).

A unique mechanism of chemotherapy that can be utilized in antiviral chemotherapy is illustrated by amantadine (16–18). The drug has no effect on the virus and none on the metabolism of the host cell, but it attaches to the surface of virus-susceptible cells and alters the interaction with influenza A virus. The result is a decreased ability of the virus to penetrate the membrane and to start the early stages of infection. Amantadine is an organic amine that is stable in its biologic and physical properties. The drug and its receptor can be removed by trypsinization without damage to the cell. The presence of amantadine does not interfere with the absorption of influenza virus onto the cell receptor. Quantitative kinetic studies indicate that adsorption proceeds at the normal rate (19). Ordinarily, the viral and cell membranes then fuse and start the process of uncoating the viral genome and transcription of the genetic information. In the presence of amantadine, those processes are delayed significantly and the virus remains on the exterior surface of the susceptible cell for a much longer time. A nonspecific effect of the drug is to decrease pinocytosis, which also may participate in reducing the infectivity of the virus. Other organic amines have been observed to exert a similar biologic effect (20,21).

The chelation of trace elements, which is another categorical mechanism for the chemical inhibition of viruses, may act by decreasing the availability of receptors for the virus on the surface of the cell. The mechanism is an unlikely one for clinical applicability; however, there are some opportunities in which drugs can gain access to secretions that bathe cells and, through the chelation of specific ions, prevent the initiation or continuation of infection. An apparent example of such an action is illustrated in Figure 16.2, in which 1.0 gm/day of oxolinic acid was given to volunteers as prophylaxis against a rhinovirus infec-

VIRUS EXCRETION

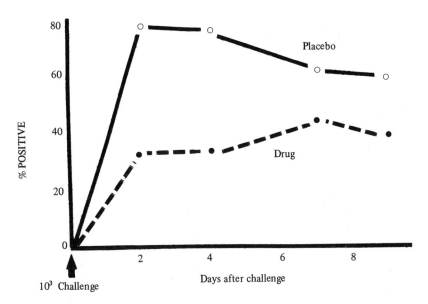

Figure 16.2. Decreased virus shedding among patients treated with 250 mg of oxolinic acid by mouth four times daily starting before a rhinovirus challenge. (From ref. 75.)

tion. The drug that chelates Mg^{++}, Mn^{++}, and Ca^{++} is secreted by lacrimal glands perhaps directly into the respiratory secretions.

The classical mechanisms of chemotherapy are those in which drugs are used to inhibit or misdirect virus-dependent metabolic processes. These involve the synthesis of structural proteins of the virus envelope, of the nucleoproteins, and of nonstructural components that are essential in the replication, assembly, or release of infectious virus (22,23). The blockade of synthetic pathways is the mode of action for most of the antiviral drugs that have received serious study or use. Among them are the halogenated pyrimidines and substituted nucleosides that, when incorporated into nucleic acids, result in noninfectious virus particles. Such is the action of idoxuridine (IDU) and cytosine arabinoside (AraC). Some nucleosides and other compounds inhibit viral DNA and RNA synthesis by stopping enzymatic action at specific steps. Adenine arabinoside (AraA), ribavirin (24), guanidine, and hydroxybenzylbenzimidazole (25) appear to have this as their primary action.

321

Some macrocyclic antibiotics like rifampins and streptovaricins inhibit RNA polymerases. Misdirected protein synthesis from abnormal RNA, or its deficiency, results in the inhibition of virus multiplication. One effect of thiosemicarbazones, which also chelate heavy metals, is to block production of two specific proteins required in the assembly of poxvirus. Still other compounds act as competitive inhibitors of critical enzyme substrates. Trifluoracetyl neuraminic acid (FANA) is such a competitive inhibitor of influenzal neuraminidase. Under some conditions, it is capable of diminishing the separation of influenza virus from infected cells (26).

A novel approach that has been investigated in the control of viral diseases is the use of drugs that have no direct antiviral effect but that stimulate natural products that inhibit viruses or increase the resistance of the host to infection or both. This is illustrated by interferon and inducers capable of eliciting its production *in vivo* (27,28). Interferon has a broad spectrum of antiviral activity, and its mechanisms of action are not entirely known but relate to the inhibition of viral-directed transfer-RNA (tRNA). Cells that are uninfected respond to interferon with the production of viral inhibitory protein or proteins that interfere with the translation of viral mRNA (29,30). Interferon and the viral inhibitory proteins have no appreciable antigenicity nor adverse effect on the cells or the immunologic response of the host.

Increased resistance is also the rationale used by those who recommend medicinal doses of vitamin C as protection against viral respiratory tract infections (31,32). The mechanism of action, if any, is unknown, but it is postulated that the lysosome-mediated damage of infected tissues is stabilized by ascorbic acid. During an acute viral infection, the concentration of ascorbic acid in leukocytes is reported to show a transient decline (33). Other drugs that directly stimulate cell-mediated immunity are being investigated for their potential in increasing host efficiency in resisting viral infections. Two for which such a mechanism is claimed are levamisol and isoprinosine.

CLINICAL INVESTIGATION AND USE

With the various metabolic antagonists it has been possible to probe viral replication in well-defined systems and to introduce effective and precise antiviral chemotherapy in tissue cultures and sometimes in in-

fected animals. Clinical application is often restricted by pharmaco-
kinetic factors, cytotoxicity, or the specificity of the action. Sometimes
the specificity differs among strains even within the same type of virus.
Nevertheless, antiviral drugs that are licensed and some that are still
investigational have had well-controlled, extensive, clinical trials.

Amantadine

Prophylaxis Against Influenza A. The administration of amantadine,
100 mg two or three times daily, before a challenge of volunteers with
influenza A virus resulted in decreased infection, restricted viral replication, and a diminished serologic response reflecting a limitation of
infection (16,34). Persons pretreated with amantadine had fewer
symptoms and less fever (35,36). Under these conditions, amantadine
was 50% to 60% effective in preventing infection. The prevention of
naturally acquired infection with influenza A was shown in four semi-
closed institutions (37). Among 601 persons treated with amantadine
who had low serum titers of antibody to influenza at the beginning of
the study, there was a 57% reduction in the number who developed infection compared with an untreated group. Of 786 persons with antibody against influenza A virus, the rate of infection was only one third
as frequent as occurred in the low-antibody-level group. Treatment
with amantadine further reduced the rate by 20%. Influenzal illness
was 62% less frequent among persons given amantadine, and one
third fewer of the treated persons who became infected had clinical influenza as compared with similar untreated controls.

A study of amantadine given prophylactically to family members of
patients with influenza A was performed in England by Galbraith and
associates (38). In 13 families with 48 persons treated with amantadine, there was a significant decrease in influenza and/or serologic evidence of asymptomatic infection when compared with 22 families
having 69 persons in a placebo-treated group (Table 16.2). The difference, which demonstrates a 61% reduction in infection and prevention of disease, is additional confirmation of the studies in volunteers
and represents practical use of chemotherapy for protection against
influenza A.

Another important prophylactic study with amantadine by O'Donoghue and colleagues in Seattle, Washington (Table 16.2) (39) was

TABLE 16.2. *Prevention of Secondary Cases of Influenza by Amantadine Prophylaxis**

Test Group	Treatment	Number Ob-served	Influenza (Clinical and Serologic)	No Influenza	
				Serologic Rise	No Infection
Families	Amantadine	48	0 (0%)	7 (15%)	41 (85%)
	Placebo	69	10 (15%)	27 (39%)	32 (46%)
Hospital	Amantadine	50	0 (0%)	2 (4%)	48 (96%)
	Placebo	61	7 (12%)	5 (8%)	49 (80%)

* Data from refs. 38 and 39.

the prevention of hospital-acquired influenza. During a one-month period when influenza was occurring in the community and among patients admitted to the hospital, no clinical influenza developed in patients treated with amantadine after their admission; new cases continued to occur in untreated patients admitted to the same wards. As in the family study, prophylactic administration of amantadine prevented disease and reduced hospital infections by 90%. We have had similar results in preventing secondary cases of influenza A in hospitalized patients and high-risk outpatients during epidemics.

In other trials, amantadine prophylaxis has been found effective against new variants of influenza A but not against influenza B or other viral causes of clinical influenza (Fig. 16.3) (40,41). Also, the beneficial effects of drug prophylaxis are not continued beyond the time of drug administration, as was shown during periods of administration and withdrawal of amantadine during an epidemic (Table 16.3) (8). It is wise, therefore, and perhaps even necessary, to vaccinate persons during the period of drug prophylaxis if that was not done previously.

Treatment of Influenza A. The therapeutic benefit of amantadine in patients with influenza is more difficult to demonstrate than is its prophylactic usefulness. Studies in volunteers failed to show significant virologic or serologic differences if treatment was delayed until after virus challenge (34). On the other hand, symptomatic improvement was observed consistently both in volunteers and in naturally infected persons when treatment was initiated within 24 hours after the onset of symptoms (7). More recently, physiologic data showing a therapeutic

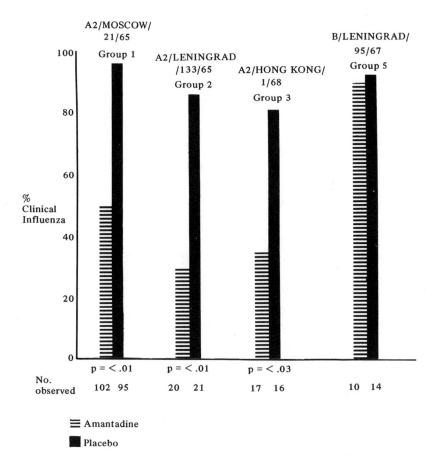

Figure 16.3. Protection against infection with recent isolates of influenza A by the prophylactic administration of amantadine and the absence of protection against influenza B. (Modified from ref. 40.)

effect of amantadine have been developed. In patients with influenza, treatment rapidly improved the increased resistance in the small airways of the lungs caused by the infection (42).

Side Effects. At the recommended dosage of amantadine, 100 mg twice each day, no important symptomatic or psychological reactions were observed in a small number of volunteers or in a large field trial (37,43). At higher doses there was clearly a dose-related amphetamine-

325

TABLE 16.3. *Relation of Influenza to Amantadine Prophylaxis and Withdrawal*

	Amantadine Given	Influenza* (Clinical and Serologic)	No Influenza	
			Serologic Rise	No Infection
Group 1 (N = 53)				
First period	Yes	2%	9%	89%
Second period	No	19%	4%	77%
Third period	No	8%	14%	78%
Group 2 (N = 52)				
First period	No	15%	4%	81%
Second period	No	5%	7%	88%
Third period	Yes	5%	0%	95%

* Percent based on uninfected persons starting each period (from ref. 8).

like stimulation of the central nervous system, and some persons became noticeably intoxicated. Thus the drug has potential toxicity, as has been emphasized, perhaps beyond its importance. The signs of intoxication are acute, completely reversible, and can be limited by dosage. In patients with Parkinson's disease, amantadine has been given continuously for years. From that experience, there is little reason to fear chronic toxicity (44).

Interferon

Interferon Inducers. For 20 years interferon has offered an exciting potential for antiviral treatment, but it presents problems in practical clinical application (27). One way to study the effects produced by interferon is to elicit its production *in vivo* by the use of chemical inducers (28). Systemic administration of potent interferon inducers has generally caused unacceptable side effects. We have studied one agent, a substituted propane diamine, that can be applied topically into the nostrils (45–47). Figure 16.4 shows the effect of topical pretreatment with the inducer in preventing rhinovirus infection and clinical illness in volunteers. When a measurable amount of interferon in the nasal secretions was stimulated before the challenge virus was given, symptoms of infection were almost entirely prevented. Less of an effect was noted on the amount of virus shed in the nasal secretion, although there was a significant reduction. When interferon was not elicited until after infection, the beneficial effect appeared to be related to the titer of interferon in the nasal secretions. After a level of 10–100 units/ml was attained in the nasal washings, symptoms declined abruptly. This occurred much before any antibody could be found in the respiratory secretions or serum. Similar results have been obtained in other trials using a different chemical inducer (48) or human interferon (49) and a challenge of either rhinovirus or influenza A or B.

Recently the effect of interferon on infection with hepatitis B virus has been reported (50). In preliminary studies, the use of an interferon inducer in chimpanzees having chronic hepatitis B virus infection caused either the elimination or a marked reduction in the number of infectious particles. Preliminary studies in patients with chronic hepatitis caused by hepatitis B using preformed human interferon also have shown suppression of Dane particles, the infectious unit of hepatitis B virus.

Preformed Interferon. Interferon probably offers more promise of clinical applicability than the use of chemical inducers owing to the rapid development of patient unresponsiveness to the latter. Treatment of human leukocytes *in vitro* permits the harvest of considerable amounts of interferon. This material has been used with promising results in the control of disseminated herpesvirus and varicella-zoster infections (51). The topical administration of interferon in the eye with or without debridement has produced minimal therapeutic effect (52). The frequency of recurrences may or may not be improved (53,54). In all these infections, interferon may be effective when given prophylactically under experimental conditions. However, as with other forms of antiviral treatment, therapeutic effects are difficult to achieve in persons with advanced infection and in patients with immunodeficiencies.

In early clinical trials with interferon and interferon inducers, a major problem always occurred in obtaining high enough levels of interferon to ensure an adequate test of its effect in different types of infections. Recently, human leukocyte interferon has been prepared in sufficient concentration and purity to administer in doses as large as 10^5–10^6 IU/kg/day (55). Even at those doses, however, the effect on viremia or viruria with CMV has been slight (56). Also, some fever and abnormal liver function tests were induced by the treatments with these large doses.

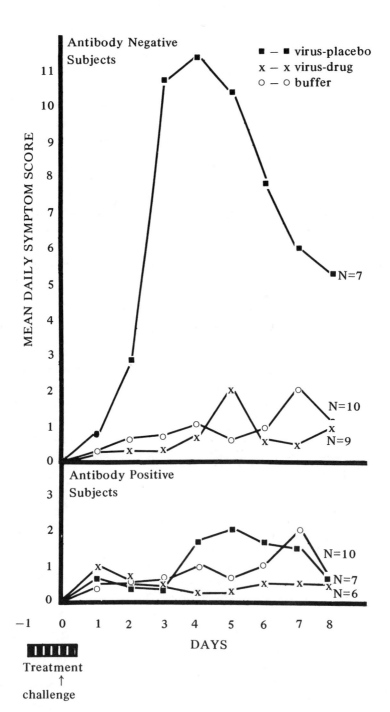

Nucleoside Analogues

Idoxuridine

HERPETIC KERATITIS. The effectiveness of IDU as topical treatment for dendritic ulcers of the cornea caused by HSV is established beyond doubt (4–6). However, such treatment has no beneficial effect on the stromal or uveal manifestations associated with herpesvirus ophthalmitis. Also, treatment neither prevents not alters the frequency of recurrences of herpetic keratitis. During treatment, IDU-resistant isolates of herpesvirus can be recovered. These isolates have been infrequently associated with clinical disease, but recent reports implicate IDU-resistant viruses with pathological changes. Newer drugs, however, may circumvent the problem (57).

HERPESVIRUS ENCEPHALITIS. In 1970 a preliminary report suggested that IDU administered intravenously in a dose of 50–80 mg/kg/day for five days increased the survival and decreased the neurological residual of patients with HSV encephalitis (58). *In vitro* studies suggested that the minimal inhibitory concentration of IDU for HSV was from 5–10 or 50–75 μg/ml, depending on the system used (58,59). Pharmacologic studies showed that the rate of infusion of IDU needed to exceed 4 mg/min to overcome the rate of inactivation from deiodination and excretion and produce a demonstrable level of antiviral activity in body fluids (60). When IDU was given to nine patients with proved or probable herpesvirus encephalitis in doses of 100 mg/kg/day, it was found to be too toxic for use (61). Seven of the nine patients died after completion of the course of therapy. HSV was isolated from the brains of all four biopsy-proved cases on whom autopsies were performed. When a basis was available for quantitative evaluation, the titers of virus before and after treatment with IDU were about equal.

Figure 16.4. The difference in symptomatic illness after rhinovirus challenge among persons treated with a topical interferon inducer and placebo-treated controls with comparable prechallenge antibody status. (From ref. 46.)

Leukopenia, thrombocytopenia, and hepatotoxicity are effects of systemic administration of IDU. Doses in excess of 50 mg/kg/day for five days produce pathological depression of platelets and neutrophils, with maximum depression encountered 10 to 14 days after the cessation of treatment. Stomatitis, diarrhea, hemorrhage, and alopecia are additional common complications. In patients who survive, these changes are all reversible.

An important lesson about antiviral chemotherapy can be learned from the experience in the treatment of herpesvirus encephalitis. Although it was uncontrolled, the initial clinical impression that treatment was beneficial had such wide acceptance that it nearly precluded the opportunity to withhold such treatment in order to conduct a controlled trial. When the additional investigational trial was conducted, the treatment was found to be detrimental. Now it is generally agreed that IDU is not useful in the treatment of herpes virus encephalitis.

VARICELLA INFECTIONS. In the use of IDU for the treatment of varicella pneumonia, disseminated varicella, generalized cutaneous varicella, and herpes zoster in 28 immunosuppressed children, new lesions ceased to appear within one or two days in all the conditions except herpes zoster (62). The drug was given in four doses per day of about 25 mg/kg per dose for five days. Deaths that occurred were from uncontrolled infection, and none was attributed to drug toxicity. The antiviral effects were uncontrolled, and the beneficial effect was clinical.

AraC and AraA. Figure 16.5 shows the structure of cytosine arabinoside (AraC) and adenine arabinoside (AraA) and a related furanosyl, all of which are undergoing clinical investigation as potential antiviral drugs for clinical use. They are grouped together because they have some chemical similarities, even though they are not comparable in their biologic effects and have different prospects as antiviral drugs. The two nucleosides, AraC and AraA, also have different biologic actions. In cell cultures, herpesviruses are inhibited by AraC at 20- to 30-fold lower concentrations than are needed with AraA (59). The former is antagonistic to IDU, while the latter is additive. In tissue cultures, the 50% plaque inhibition concentration for HSV is 310 μg/ml of AraA. After infusion, AraA is rapidly converted to the hypoxanthine (AraHx), which has only 10% to 50% of the antiviral activity

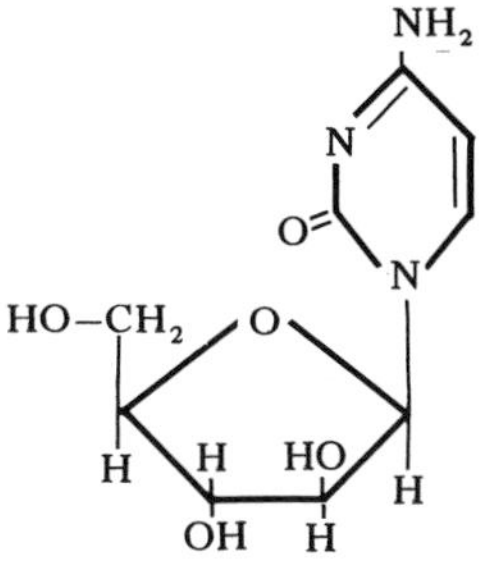

1-β-D-arabinofuranosyl cytosine (ara-C)

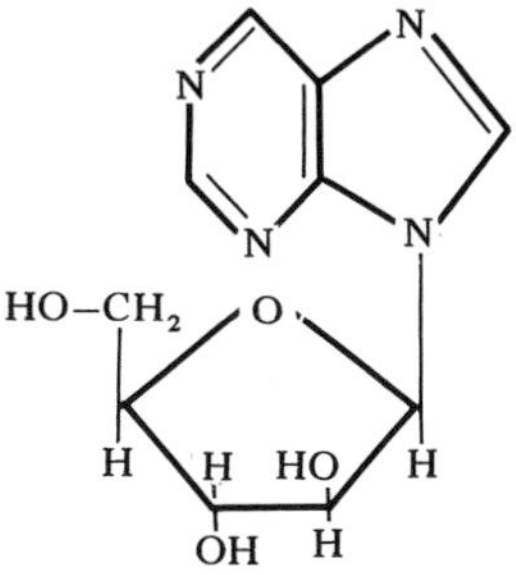

9-β-D-arabinofuranosyladenine (ara-A)

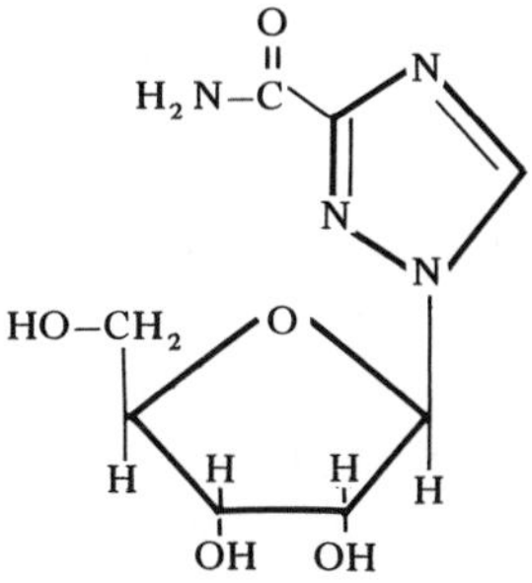

1-β-D-ribofuranosyl-1,2,4-triazole-3-carboxamide
(ribavirin-Virazole®)

Figure 16.5.   Three chemically related antagonists of viral nucleic acid synthesis that have antiviral activity and are undergoing clinical investigation.

of the parent compound. When 10 mg/kg is given every four to six hours, it produces a plasma level of 6–8 μg/ml of AraHx. At that dose, toxic manifestations of AraA have not been the limiting factor they are with AraC and IDU (63). Nausea, vomiting, tremors, and toxic encephalopathy have been clinical side effects in some patients, especially at higher doses. A variable degree of megaloblastic maturation arrest of erythropoiesis is detectable after several consecutive days of treatment with AraA, but thrombocytopenia is a minor risk compared with the effects of AraC and IDU.

VIRAL ENCEPHALITIS. As with IDU, AraC requires intolerable doses for the treatment of CNS infections, and its use is not recommended. AraA, on the other hand, has antiviral activity in a dose range that has caused no apparent toxicity in some studies. Its use in neonates with HSV encephalitis in dosages of 10–20 mg/kg/day has been successful in a few patients when the drug was started within three days of the first signs of CNS involvement. Treatment started later was unsuccessful (64). On the basis of this experience, controlled trials are in progress.

VARICELLA-ZOSTER. "Shingles" and disseminated herpetic cutaneous lesions tend to occur in patients with impairment of the immune system, but sometimes there is no apparent immunologic defect in the host (65). With the variability of host conditions, the course of the disease is extremely difficult to predict. An apparently beneficial effect of different treatment regimens has misled seasoned investigators. Some reports suggest a beneficial effect of AraC on disseminated herpes zoster. However, in a randomized double-blind evaluation of 19 treated and 20 control patients, in whom treatment with 100 mg/m²/day was started within the first 48 hours and given by continuous intravenous infusion for up to 72 hours, the effect of the drug was paradoxical (66). The treated group showed a more prolonged period of dissemination than the placebo group. The adversity was believed to be mediated by suppression of antibody response, decreased cellular immune response, and a delay in the appearance of interferon in the vesicular fluid.

Results in the treatment of herpes zoster with AraA are more encouraging. In an initial collaborative trial, 87 patients were treated alternately with AraA or placebo in sequential five-day periods. The dosage of AraA given was 10 mg/kg/day. During the first five days, the resolution of skin lesions and pain was significantly faster in the treated group (67,68).

Ribavirin

PREVENTION OF INFLUENZA. Ribavirin inhibits both influenza A and B viruses as well as other viruses that cause respiratory diseases (69) and has a direct antiviral action. In tissue cultures, 0.1 μg/ml inhibits replication of influenza A, and an oral dose of 400 mg of ribavirin produced a plasma level of 0.2 μg/ml.

Initial trials in volunteers using a dosage of 600 mg/day produced a marginal effect against influenza B and none against influenza A (70, 71). Preliminary reports using a higher dosage indicate that anti-influenzal activity can also be demonstrated against influenza A. In an outbreak of influenza A among girls aged 8 to 16 years in a boarding school in Mexico, treatment with 100 mg of ribavirin three times daily was reported to be more than 80% effective in preventing disease (72).

HEPATITIS. In several different placebo-controlled studies in Brazil that included 72 patients diagnosed as having hepatitis A, a dose of 100 mg of ribavirin given four times daily accelerated the return of abnormal elevations of liver enzymes in the serum and of the serum bilirubin (73,74). A similar study of patients with hepatitis B in the United States showed no beneficial effect.

The toxicity of short courses with the above dosages has been minimal, but the toxicity at higher levels that produce more significant antiviral activity *in vivo* must still be determined. It is known from studies in animals that ribavirin can produce anemia, immunosuppression, and teratogenesis (in rodents).

MISCELLANEOUS CLINICAL TRIALS

Respiratory Infections

Several compounds have been shown to have selective antiviral activity in tissue culture against different picornaviruses that cause respiratory infection (75–78). Some cause little or no toxicity to the tissue cells at antiviral concentrations. A few have had preliminary trials in volunteers given a rhinovirus challenge (75,79–81). Figure 16.6 shows the results with one such drug, a substituted triazine indole. When it was administered orally beginning at the time of the challenge, the prophylactic effect in a double-blind placebo-controlled study was significant in preventing illness from infection. Among comparable subjects, the rate was reduced by one half. Unfortunately, two features that commonly plague effective antiviral chemotherapy were observed—the relation of effectiveness to the time of drug administration and symptomatic side effects from the treatment. More than three fourths of the drug effect was lost when it was given as therapy, and about 10% of treated persons had severe headache from the drug. Thus, further trials were not conducted, although the results demonstrated the feasibility of chemical interference with a mucosal infection by the oral administration of a drug that inhibits viral multiplication.

Figure 16.6. The prevention of clinical illness after a rhinovirus challenge by the prophylactic oral administration of an investigational drug (a substituted triazine indole) capable of inhibiting the synthesis of rhinovirus. The therapeutic effect was insignificant, and side effects from the drug occurred.

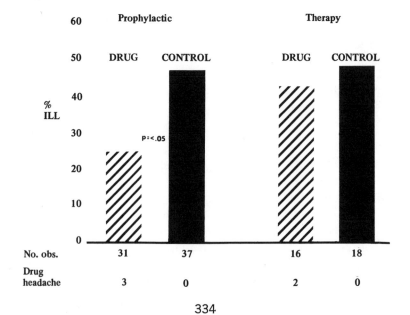

The report of 39% decreased frequency of colds at a ski school in Switzerland among 139 of 279 students after the regular ingestion of 1 gm daily of vitamin C and the much publicized book of an eminent biophysical scientist stimulated interest and some trials of this method of prophylaxis or therapy for the common cold (31,32). Most of the controlled epidemiologic studies have failed to show a significant or important difference in the incidence, severity, or duration of respiratory illnesses from such treatment (82). A few have shown significant differences in some features of the illness or disability (83,84). Usually there is a suspicion that these are personal or sociologic differences; no one has shown the treatment to have an antiviral effect. Trials in volunteers have failed to provide any support for the role of ascorbic acid in the clinical or virologic aspects of these diseases (85,86).

SUMMARY AND PERSPECTIVE

In the development of antiviral chemotherapy, an even and objective attitude is necessary. The desire to succeed is tremendous. Testimonials of success can be convincing, and the literature is full of enthusiastic initial claims and rational postulates that became dashed by experience. On the other hand, philosophic pessimists who unreasonably doubt the feasibility of practical antiviral chemotherapy still abound. Although they have been slower and more marginal than desired, clear-cut successes and tremendous progress toward practical antiviral chemotherapy have been achieved. The discovery and development of more new mechanisms of chemotherapy resulting from the peculiar nature of viral infections can be expected and should provide new intellectual stimulation.

Barring a serendipitous discovery, continued progress will require continued improvement in our basic understanding of viral diseases. More virologic, clinical, and epidemiologic investigations of viral diseases are needed. The magnitude of the ignorance and abuse in the clinical designation of viral diseases, notions about their pathogenesis, and considerations of their course can hardly be imagined. All manner of statements and perceptions about possible remedies are thus permitted. The attempts to develop effective chemotherapy keep reemphasizing our dependence on an improved ability to specifically recognize and predict the course of the viral diseases we aim to treat.

Besides the difficulties imposed by the intimacy of viral replication and cellular metabolism, the great importance of the time of treatment on the outcome obtained by chemotherapy has already been demonstrated repeatedly. In the achievement of effective use of drugs in viral infections, practices will probably need to move progressively toward prophylaxis and early treatment. Both are dependent of diagnostic capabilities that are rarely available. The goal for later treatment may need to be the confinement of infection—either within the patient or in epidemiologic terms such as "drug quarantine" of infected cases—and the prevention of transmission. The use of prevention as a measure of drug efficacy makes objective evaluation even more difficult. Because some of these problems cannot be solved by clinical trials, appropriate model systems must continue to be developed. Interpretable infections in animals and, where ethical, controlled tests in volunteers and defined populations will be important, if not absolutely necessary, to place antiviral chemotherapy on a firm basis. The sociologic acceptance of controlled trials and the laborious requirements of their conduct must be promulgated until specific treatments are confirmed with the range of their benefits and restrictions. Chronic viral infections and those with exacerbations related to the immunologic status of the host are particular clinical problems that are more complex than just antiviral chemotherapy.

The potential for toxicity of most of the antiviral drugs will require an increasing effort to learn the quantitative pharmacokinetics of the drugs. Even with the drugs licensed for clinical use and those that have had much clinical investigation, too little is known about the optimum dosage or drug level, the best frequency of dosage, and quantitative correlates of effective and ineffective treatment. Examples of paradoxical effects such as those observed with IDU in herpesvirus encephalitis and AraC in herpes zoster are not uncommon. The appreciation that drugs can have an enhancing effect on the viral infection or be detrimental to the host mechanisms involved in its control is important (66, 75,87).

Finally, education in the prescription and use of antiviral drugs may be equally as difficult as the biologic problems. The difficulties reflect all the burdens of too little precision in diagnosis, late initiation of treatment, unpredictability of the course, use of empiric doses, and sometimes marginal therapeutic to toxic ratios of the drugs. But unfamiliarity and inexperience with antiviral chemotherapy may be primary factors. Ten years after its licensure, amantadine still has not obtained the consideration and use that would appear justified from the record established in controlled clinical investigations. Preconceived bias, unavailability or underuse of immediate health care, nonaccess to the drug, and the patients' noncompliance with recommendations are matters that impinge on practical antiviral chemotherapy and are subject to teaching and experience. Each of the most promising new antiviral drugs has specificities, limitations, and potential toxicity that will require detailed education of the users and prescribers.

The road to effective antiviral chemotherapy appears to be stretched out well ahead of us. The directional signs are clear, and the supporting structures must be methodically developed if we are to continue to progress by means other than trial and error. Along with the discovery of antiviral chemicals with clinical potential, our goals include improved precision in clinical virology with earlier specific diagnosis and treatment, better determination of drug pharmacology in man, the development of attitudes and clinical methods for objective evaluation of treatment, and education gained from experience with antiviral chemotherapy where available.

DISCUSSION

Question: Would you please comment on the use of (a) vitamin C for the common cold, and (b) the use of steroids for infectious mononucleosis and "mycins" for influenza.

Dr. Jackson: I don't know any "mycin" that has any effect on influenza. In infectious mononucleosis with the exception of some patients with very severe agranulocytic angina or an equivalent condition, the steroidal effect on the viral infection per se is antagonistic, and the benefit that is gained in symptomatic improvement, in my view, is not worth the use of steroids. We do not use them unless the patient has a severe symptomatic complaint, and then we do so with hesitation.

The vitamin C question is a lot of fun, but it has also diverted a tremendous amount of attention and investigational time and some resources into what, I think, is a nonsensical area because negative data are difficult to use in proof. One always has to use a bigger dose or some other kinds of circumstances. In trying to improve resistance to infection, one should know that some compounds (isoprinosine, leva-misole, and some others) increase cell-mediated immunity. The only rationale for using vitamin C is that, during a viral respiratory tract infection, the concentration of ascorbic acid reducing substances in the buffy-coat leukocytes decreases from about 20 $\mu g/100 \times 10^6$ cells to about half that amount. This decrease can be remedied to some extent by administration of large doses of vitamin C. The significance of this observation is unknown. So one looks, then, at what I discussed before: How do you control this? Two reasonably controlled studies suggest that if you increase the dose of vitamin C abruptly at the onset of symptoms, clinical improvement occurs more rapidly. If you continue the same dose, whether it be 1 gm/day or 6 gm/day of vitamin C throughout a month or throughout a winter, the effect does not continue. One interpretation would be that if you abruptly increase the amount of vitamin C at the onset of viral infection, you might reverse the lowering of ascorbic acid content of the leukocytes. In one epidemiologic study, some amelioration of the disease may have occurred. By and large, I think most of the studies are nonsense and not well controlled, and an equal number show no effect when the design includes a control group.

Question: What is aspirin's use in relation to its effect on the common cold?

Dr. Jackson: The recent article in *JAMA* by Dr. Stanley and her colleagues (87) deserves comment. The first thing one is always asked about in antiviral drug studies of respiratory tract infections is, What does the aspirin control group show? So we studied the aspirin control group for the first time and noted an increased virus shedding with little amelioration of symptoms. Because we thought that result could have been a statistical freak, we repeated the study and had essentially superimposable results. Therefore, at least under conditions of the investigation wherein volunteers were challenged with rhinovirus and beginning on the day of challenge, subjects were treated with 0.6 gm of aspirin three times daily for five days, the amount of virus recovered from daily washings of the noses of treated persons increased significantly as compared with the amounts recovered from the placebo-treated groups. I think that was an important finding. As you might guess, when it was published, all kinds of extrapolations of the data were made. We did speculate that the person who was excreting more virus would also perhaps be more dangerous to the community as a spreader. Our study didn't show that, but it is logical and must be put to test.

We gave the aspirin on a regular schedule for five days, not as an occasional dose. But, to me, the importance is that the effect represents a biologic principle, and it can be shown with other drugs. When you manipulate cell/virus interrelations, you can go both ways—more virus or less virus. The best interpretation, one having some data support, is that aspirin decreases the T-cell responsiveness. Some data show that rhinovirus-sensitized T cells are effective in controlling the clinical and virologic aspects of the disease. By all evidence, I don't see any reason to use aspirin unless you have severe illness.

Dr. Sanders: Your publication was of interest to Dr. Bruce Burlington and me. Dr. Burlington, working in my laboratory, showed that aspirin in potentially toxic concentrations of 35–40 mg/dl significantly suppressed leukocyte interferon levels, that is, leukocyte interferon induced by live Newcastle disease virus.

REFERENCES

1.   Bauer DJ, St. Vincent L, Kempe CH, et al: Prophylactic treatment of smallpox with N-methyl-isatin B thiosemicarbazone (compound 33 T 57). *Lancet* 2:494–496, 1963.
2.   Rao AR, McFadzean JA, Kamalakashi S: An isothiazole thiosemicarbazone in the treatment of variola major in man. *Lancet* 1:1068–1072, 1965.
3.   Rao AR, McKendrick GDW, Belayaduhan L, et al: Assessment of an isothiazole thiosemicarbazone in the prophylaxis of contacts of variola major. *Lancet* 1:1072–1074, 1966.
4.   Kaufman HE: Clinical cure of herpes simplex keratitis by 5-iodo-2'-deoxyuridine. *Proc Soc Exp Biol Med* 109:251–252, 1962.
5.   Laibson PR, Leopold IH: An evaluation of double-blind IDU therapy in one hundred cases of herpetic keratitis. *Trans Am Acad Ophthalmol Otolaryngol* 68:22–24, 1964.
6.   Gold JA, Stewart RC, McKee J: The epidemiology and chemotherapy of herpes simplex keratitis and herpes simplex skin infections. *Ann NY Acad Sci* 130:209–212, 1965.
7.   Jackson GG, Stanley ED: Prevention and control of influenza by chemoprophylaxis and chemotherapy. *JAMA* 235:2739–2742, 1976.
8.   Muldoon RL, Stanley ED, Jackson GG: Use and withdrawal of amantadine chemoprophylaxis during epidemic influenza A. *Am Rev Respir Dis* 113:487–491, 1976.

9. Wallis C, Melnick JL: Irreversible photosensitization of viruses. *Virology* 23:520–527, 1964.

10. Rapp F, Li JH, Jerkofsky M: Transformation of mammalian cells by DNA-containing viruses following photodynamic inactivation. *Virology* 55: 339–346, 1973.

11. Myers MG, Oxman MN, Clark JE, et al: Photodynamic inactivation in recurrent infections with herpes simplex virus. *J Infect Dis* 113(S):145–150, 1976.

12. Engle CG, Liu OC: Studies on the chemotherapy of viral infections: III. Antiinfluenza activities of glyoxal analogues *in vitro* and in chick embryo systems. *J Immunol* 89:531–538, 1962.

13. Jackson GG, Muldoon RL, Akers LW, et al: Effect of N′N′-anhydrobis (B-hydroxyethyl) biguanide hydrochloride on influenza virus in volunteers. *Antimicrob Agents Chemother* 1963, pp 883–891.

14. Williamson GM, Jackson D: The antiviral activity of the isoquinolines famotine and memotine in respiratory infections in man. *Bull WHO* 41: 665–670, 1969.

15. Couch RB, Jackson GG: Antiviral agents in influenza—summary of influenza workshop VIII. *J Infect Dis* 134:516–527, 1976.

16. Jackson GG, Muldoon RL, Akers LW: Serological evidence for prevention of influenzal infection in volunteers by an antiinfluenzal drug, Amantadine hydrochloride. *Antimicrob Agents Chemother* 1963, pp 703–707.

17. Davies WL, Grunert RR, Haff RL, et al: Antiviral activity of 1-adamantanamine (amantadine). *Science* 144:862–863, 1964.

18. Hoffman CE, Newmeyer EM, Hoff RF, et al: Mode of action of the antiviral activity of amantadine in tissue culture. *J Bacteriol* 90:623–628, 1965.

19. Kato N, Eggers J: Inhibition of uncoating of fowl plaque virus by 1-adamantanamine hydrochloride. *Virology* 37:632–641, 1969.

20. Fletcher RD, Hirschfield JE, Forbes M: A common mode of antiviral action for ammonium ions and various amines. *Nature* 207:664–665, 1965.

21. Togo Y, Schwartz AR, Tominaga S, et al: Cyclooctylamine in the prevention of experimental human influenza. *JAMA* 220:837–841, 1972.

22. Carter WA (ed): *Selective Inhibitors of Viral Functions.* Cleveland, CRC Press, 1973.

23. Becker Y: *Antiviral Drugs: Mode of Action and Chemotherapy of Viral Infections of Man.* Monographs in Virology. JL Melnick (ed), Basel, S Karger, 1976.

24. Scholtissek C: Inhibition of influenza RNA synthesis by Virazole (ribavirin). *Arch Virol* 50:349–352, 1976.

25. Caliguiri LA, Tamm I: Guanidine and 2(α-Hydroxybenzyl)-Benzimidazole (HBB): Selective inhibitors of picornaviruses multiplication, in Carter WA (ed): *Selective Inhibitors of Viral Functions.* Cleveland, CRC Press, 1973, pp 257–293.

26. Schulman JL, Palese P: Susceptibility of different strains of influenza A virus to the inhibitory effects of 2-deoxy-2,3-dehydro-N-trifluoroacetylneuraminic. *Virology* 63:98–104, 1975.

27. Finter NB: Exogenous interferon in animals and its clinical implications. *Arch Intern Med* 126:147–157, 1970.

28. DeClercq E, Merigan TC: Induction of interferon by nonviral agent. *Arch Intern Med* 126:94–108, 1970.

29. Joklik WK, Merigan TC: Concerning the mechanism of action of interferon. *Proc Natl Acad Sci* 56:558–565, 1966.

30. Marcus PI, Salb JM: Molecular basis of interferon action: Inhibition of viral RNA translation. *Virology* 30:502–516, 1966.

31. Ritzel G: Kritische Beurteilung des vitamin C als prophylacticum und therapeuticum der Erkaltungs krankheiten. *Helv Med Acta* 28:63–68, 1961.

32. Pauling L: *Vitamin C and the Common Cold.* San Francisco, WH Freeman, 1970.

33. Hume R, Weyers E: Changes in leukocyte ascorbic acid during the common cold. *Scott Med J* 18:3–7, 1973.

34. Stanley ED, Muldoon RL, Akers LW, et al: Evaluation of antiviral drugs: The effect of amantadine on influenza in volunteers. *Ann NY Acad Sci* 130:44–51, 1965.

35. Togo Y, Hornick RB, Dawkins AT Jr: Studies on induced influenza in man: I. Double-blind studies designed to assess prophylactic efficacy of amantadine hydrochloride against A2/Rochville/1/65 strain. *JAMA* 203:1089–1094, 1968.

36. Dawkins AT Jr, Gallager LR, Togo Y, et al: Studies on induced influenza in man: II. Double-blind study designed to assess the prophylactic efficacy of an analogue of amantadine hydrochloride. *JAMA* 203:1095–1099, 1968.

37. Jackson GG, Stanley ED, Muldoon RL: *Chemoprophylaxis of Viral Respiratory Diseases.* First International Conference on Vaccines Against Viral and Rickettsial Diseases of Man. Washington DC, Pan Am Health Organization, 1967, pp 595–603.

38. Galbraith AW, Oxford JS, Schild GC, et al: Study of 1-adamantamine hydrochloride used prophylactically during the Hong Kong influenza epidemic in the family environment. *Bull WHO* 41:677, 1969.

39. O'Donoghue JM, Ray CG, Terry DW Jr, et al: Prevention of nosocomial influenzal infection with amantadine. *Am J Epidemiol* 97:276–282, 1973.

40. Smorodintsev AA, Zlydnikov DM, Kiseleva AM, et al: Evaluation of amantadine in artifically induced A2 and B influenza. *JAMA* 213:1448–1454, 1970.

41. Smorodintsev AA, Karpuchin GI, Zlydnikov DM, et al: The prospect of amantadine for prevention of influenza A2 in humans (effectiveness of amantadine during influenza A2/Hong Kong epidemics in January–February, 1969, in Leningrad). *Ann NY Acad Sci* 173:44–61, 1970.

42. Little JW, Hall WJ, Douglas RG Jr, et al: Amantadine effect on peripheral airways abnormalities in influenza. *Ann Intern Med* 85:177–182, 1976.

43. Peckinpaugh RO, Askin FB, Pierce WE, et al: Field studies with amantadine: Acceptability and protection. *Ann NY Acad Sci* 173:62–73, 1970.

44. Butzer JF, Silver DE, Sahs AL: Amantadine in Parkinson's disease. *Neurology (Minneap)* 25:603–606, 1975.

45. Gatmaitan BG, Stanley ED, Jackson GG: The limited effect of nasal

interferon induced by rhinovirus and a topical chemical inducer on the course of infection. *J Infect Dis* 127:401–407, 1973.

46. Panusarn C, Stanley ED, Dirda V, et al: Prevention of rhinovirus illness by a topical interferon inducer. *N Engl J Med* 291:57–61, 1974.

47. Stanley ED, Jackson GG, Dirda VA, et al: Effect of a topical interferon inducer on rhinovirus infections in volunteers. *J Infect Dis* 133:121–127, 1976.

48. Hill DA, Baron S, Perkins JC, et al: Evaluation of an interferon inducer in viral respiratory disease. *JAMA* 219:1179–1184, 1972.

49. Merigan TC, Hall TS, Reed SE, et al: Inhibition of respiratory virus infection by locally applied interferon. *Lancet*, 1:563–567, 1973.

50. Purcell RH, London WT, McAuliffe VJ, et al: Modification of chronic heptitis-B virus infection in chimpanzees by administration of an interferon inducer. *Lancet* 2:757–761, 1976.

51. Jordan GW, Freed R, Merigan TC: Administration of human leukocyte interferon in herpes zoster: I. Safety, circulating antiviral activity and host response to interferon. *J Infect Dis* 130:56–62, 1974.

52. Sundmacher R, Neumann-Haefelin D, Manthey KF, et al: Interferon in treatment of dendritic keratitis in humans: A preliminary report. *J Infect Dis* 133:160–164, 1976.

53. Kaufman HE, Meyer RF, Laibson PR, et al: Human leukocyte interferon for the prevention of recurrences of herpetic keratitis. *J Infect Dis* 133:165–168, 1976.

54. Jones BR, Coster DJ, Falcon MG, et al: Clinical trials of topical interferon therapy and ulcerative viral keratitis. *J Infect Dis* 133:A169–172, 1976.

55. Cantell K, Hirvonen S, Mogensen KE, et al: Human leukocytic interferon: Production, purification stability and animal experiments, in: *The Production and Use of Interferon for the Treatment and Prevention of Human Virus Infections*. Proceedings of the Tissue Culture Association. Rockville, Md., Warren R Stinebring, 1974.

56. Arvin AM, Yeager AS, Merigan TC: Effect of leukocyte interferon on urinary excretion of cytomegalovirus by infants. *J Infect Dis* 133:205–210, 1976.

57. Nesburn AB, Robinson C, Dickinson R: Adenine arabinoside effect on experimental idoxuridine resistant herpes simplex infection. *Invest Ophthalmol* 13:302–304, 1974.

58. Nolan DC, Carruthers MM, Lerner AM: Herpesvirus hominis encephalitis in Michigan: Report of 13 cases, including six treated with idoxuridine. *N Engl J Med* 282:10–13, 1970.

59. Fiala M, Chow EW, Miyasaki K, et al: Susceptibility of herpesviruses to three nucleoside analogues and their combinations and enhancement of antiviral effect at acid pH. *J Infect Dis* 29:82–85, 1974.

60. Lerner AM, Bailey EJ: Concentration of idoxuridine in serum, urine, and cerebrospinal fluid of patients with suspected diagnoses of herpesvirus hominis encephalitis. *J Clin Invest* 51:45–49, 1972.

61. Boston Interhospital Virus Study Group. NIAID Sponsored Cooperative Antiviral Clinical Study: Failure of high dose 5-iodo-2'-deoxyuridine in the therapy of herpes simplex virus encephalitis: Evidence of unacceptable toxicity. *N Engl J Med* 292:599–603, 1975.

62. Feldman S, Hughes WT, Chandhary S: Antiviral therapy for varicella

(VZV) infection in children with cancer: Evaluation of idoxuridine (IDUR) (abstract #99). Sixteenth Interscience Conference on Antimicrobials, Agents & Chemotherapy, 1976.

63.   Ross AH, Julia A, Balakrishman C: Toxicity of adenine arabinoside in humans. *J Infect Dis* 133:192–198, 1976.

64.   Ch'ien LT, Whitley RJ, Nahmias AJ, et al: Antiviral chemotherapy and neonatal herpes simplex virus infection: A pilot study—experience with adenine arabinoside (Ara-A). *Pediatrics* 55:678–685, 1975.

65.   Naraqi S, Jackson GG, Jonasson O: Viremia with herpes simplex type I in adults: Four nonfatal cases, one with features of chicken pox. *Ann Intern Med* 85:165–169, 1976.

66.   Stevens DA, Jorday GW, Waddell TF, et al: Adverse effect of cytosine arabinoside on disseminated zoster in a controlled trial. *N Engl J Med* 289:873–878, 1973.

67.   Ch'ien LT, Whitley RJ, Alford CA Jr, et al: Adenine arabinoside for therapy of herpes zoster in immunosuppressed patients: Preliminary results of a collaborative study. *J Infect Dis* 133:184–191, 1976.

68.   Whitley RJ, Ch'ien LT, Dolin R, et al: Adenine arabinoside therapy of herpes zoster in the immunosuppressed, NIAID collaborative antiviral study. *N Engl J Med* 294:1193–1199, 1976.

69.   Sidwell RW, Huffman JH, Khare GB, et al: Broad spectrum antiviral activity of virazole, 1-B-D ribofuranosyl-1,2,4-triazole-3-carboxamide. *Science* 177:705–706, 1972.

70.   Cohen A, Togo Y, Khakhoo R, et al: Comparative clinical and laboratory evaluation of the prophylactic capacity of ribavirin, amantadine hydrochloride, and placebo in induced human influenza type A. *J Infect Dis* 133: 114–120, 1976.

71.   Togo Y, McCracken EA: Double-blind clinical assessment of ribavirin (virazole) in the prevention of induced infection with type B influenza virus. *J Infect Dis* 133:109–113, 1976.

72.   Salido-Rengell F, Nasser-Quinones H, Briseno-Garcia B: Clinical evaluation of 1-β-D-ribofuranosyl-1,2,4-triazole-3-carboxamide (ribavirin) in a double-blind study during an outbreak of influenza. *Ann NY Acad Sci* 284:272–277, 1977.

73.   Zuniga CB, de Almeida C, Ierovolino ACL, et al: Action of 1-β-D ribofuranosyl, 1,2,4 triazole-3-carboxamide (Viramid, ICN-1229) in the treatment of acute viral hepatitis. *Rev Assoc Med Bras* 20:385–390, 1974.

74.   Galvao PAA, Castro IO: Treatment of acute viral hepatitis with a new antiviral compound. *Rev Bras Clin Terapeut* 3:220–228, 1965.

75.   Jackson GG: A perspective from controlled investigations on chemotherapy for viral respiratory infections. *J Infect Dis* 133:83–92, 1976.

76.   Gwaltney JMP: Rhinovirus inhibition by three substituted triazine indoles. *Proc Soc Exp Biol Med* 133:1148–1154, 1970.

77.   Reed SE, Bynoe ML: The antiviral activity of isoquinoline drugs for rhinoviruses in vitro and in vivo. *J Med Microbiol* 3:346–352, 1970.

78.   Reed SE, Craig JW, Tyrrell DAJ: Four compounds active against rhinovirus: Comparisons in vitro and in volunteers. *J Infect Dis* 133:128–135, 1976.

79.   Togo Y, Schwartz AR, Hornick RB: Antiviral effect of 3,4-dihydro-1-isoquinolineacetamide hydrochloride in experimental human rhinovirus infection. *Antimicrob Agents Chemother* 4:612–616, 1973.

80. Soto AJ, Hall TS, Reed SE: Trial of the antiviral action of isoprinosine against rhinovirus infection of volunteers. *Antimicrob Agents Chemother* 3:332–334, 1973.

81. Pachuta DM, Togo Y, Hornick RB, et al: Evaluation of isoprinosine in experimental human rhinovirus infection. *Antimicrob Agents Chemother* 5:403–408, 1974.

82. Karlowski TR, Chalmers TC, Frenkel LD, et al: Ascorbic acid for the common cold: A prophylactic and therapeutic trial. *JAMA* 231:1038–1042, 1975.

83. Andersen TW, Reid DBW, Beaton GH: Vitamin C and the common cold: A double-blind trial. *Can Med Assoc J* 107:503–508, 1972.

84. Coulehan JL, Reisinger KS, Rogers KD, et al: Vitamin C prophylaxis in a boarding school. *N Engl J Med* 290:6–10, 1974.

85. Walker GH, Bynoe ML, Tyrrell DAJ: Trial of ascorbic acid in prevention of colds. *Br Med J* 1:603–607, 1967.

86. Schwartz AR, Togo Y, Hornick RB, et al: Evaluation of the efficacy of ascorbic acid in prophylaxis of induced rhinovirus 44 infection in man. *J Infect Dis* 128:500–505, 1973.

87. Stanley ED, Jackson GG, Panusarn C, et al: Increased virus shedding with aspirin treatment of rhinovirus infection. *JAMA* 231:1248–1251, 1975.

Abbreviations

ABOB	N'N'-anhydrobis (B-hydroxyethyl) biguanide hydrochloride
AKR	an inbred strain of mice
AraA	adenine arabinoside
AraC	cytosine arabinoside
AraHx	hypoxanthine arabinoside
ARD	acute respiratory disease
BKV	a human papovavirus
BrdU	5-bromodeoxyuridine
cDNA	complementary DNA
CE	chicken embryo
CF	complement fixation
Chla-Str	chlamydozoa-strongyloplasm
CJD	Creutzfeldt-Jakob disease
CMV	cytomegalovirus
CNS	central nervous system
CPE	cytopathic effect
CsCl	cesium chloride
CSF	cerebrospinal fluid
DLE	disseminated lupus erythematosus
DMSO	dimethyl sulfoxide
DNA	deoxyribonucleic acid
EBV	Epstein-Barr herpesvirus
FANA	trifluoracetyl neuraminic acid
HAA	hepatitis-associated antigen
HAV	hepatitis A virus
HBcAg	hepatitis B core antigen
HBeAg	hepatitis B e antigen
HBIG	high-titer hepatitis B immune serum globulin
HBsAg	hepatitis B surface antigen
HBV	hepatitis B virus
HI	hemagglutination inhibition
HSV	herpes simplex virus
HSV-1	herpes simplex virus type 1
HSV-2	herpes simplex virus type 2
ICNV	International Committee on Nomenclature of Viruses
ICTV	International Committee on Taxonomy of Viruses
IDU	idoxuridine; 5-iodo-2' deoxyuridine
IPV	inactivated poliomyelitis vaccine
ISG	immune serum globulin
JCV	a human papovavirus
LCM	lymphocytic choriomeningitis
MAC	monkey-adapting compound
MHV	mouse hepatitis virus
MMTV	mouse mammary tumor virus
MoMTV	monkey mammary tumor virus (previously termed Mason-Pfizer virus)
MP	designation for one of several simian viruses
mRNA	messenger ribonucleic acid
MVM	minute virus of mice
MW	molecular weight
NT	neutralizing antibody
OPV	oral poliomyelitis vaccine
PML	progressive multifocal leukoencephalopathy
PPD	purified protein derivative, the antigen(s) used in the skin test for hypersensitivity to mycobacteria
RLV	Rauscher leukemia virus

ABBREVIATIONS

RNA	ribonucleic acid	SSV	simian sarcoma virus
RSSE	Russian spring-summer encephalitis	SV40	simian virus 40
		T antigen	tumor antigen
RSV	Rous sarcoma virus; respiratory syncytial virus	TME	transmissible mink encephalopathy
SGOT	serum glutamic oxalo-acetic transaminase	TMV	tobacco mosaic virus
		tRNA	transfer ribonucleic acid
SLE	systemic lupus erythema-tosus	TSTA	tumor-specific transplan-tation antigens
SLEV	St. Louis encephalitis virus	VZV	varicella-zoster virus
SSPE	subacute sclerosing panencephalitis	WM	Woolly monkey

Glossary

anamnestic response: pertaining to the accelerated and enhanced synthesis of antibody in response to an antigen to which the subject has previously developed a primary immune response; called also *recall, booster,* and *secondary immune response.*

attenuated: pertaining to reduced virulence, as, e.g., live poliomyelitis vaccine.

A-type, B-type, or *C-type virus:* the morphological classification by electron microscopy of RNA tumor viruses (retroviruses or oncornaviruses).

autochthonous: arising from or pertaining to self, e.g., transplantation of an animal's skin from one site to another.

B cells: bone-marrow-dependent lymphocytes that on specific interaction with antigen undergo blastogenesis and differentiate into plasma cells, the latter producing immunoglobulins specifically directed against the antigen.

capsid: the outer protein shell of a viral particle in close association with nucleic acid. In viruses having cubic symmetry, the shell is made up of capsomeres (quod vide); in viruses displaying helical symmetry, protein is bound to viral nucleic acid inside the viral envelope.

capsomere: the morphological protein subunit of a viral capsid (q.v.).

cistron: the smallest functional unit or grouping of nucleotides that must be intact to function as a transmitter of genetic information. The gene, as traditionally conceived, is identical to the cistron.

clone: the progeny of a single cell or microorganism.

complementation: the restoration of the functional capacity of an organism as a result of the mutual interaction of genes in which two or more distinct mutations have occurred, usually on distinct cistrons (q.v.), and which, therefore, alone would be inadequate.

dalton: the unit used to express molecular weight. It is equal to the mass of the hydrogen atom (1.67×10^{-24} gm).

defective particle: an incomplete (i.e., noninfectious) viral particle.

denaturation: the loss of the natural structure and/or function of a macromolecule. For example, heat treatment may cause double-stranded (q.v.) DNA to dissociate into two separate strands of DNA.

derepress: to allow a gene to be transcribed (q.v.) by release from a nonfunctioning (repressed) state.

dimer: a complex comprising two identical subunits or monomers.

double or single strand: the numerical designation of the configuration of

347

covalently linked nucleotides (DNA or RNA), either as one thread of bases or two intertwined strands of base pairs. If double-stranded, the strands are complementary in adjacent purine and pyrimidine base sequence, guanine pairing with cytosine, and adenine with thymidine or uracil.

"early" or "late" (genes): viral genes coding for viral mRNA transcribed (q.v.) before (early) or after (late) viral DNA replication begins. "Early" and "late" may also be applied to the products of these genes.

envelope: the viral coat that surrounds the capsid and contains virus-coded proteins and host-derived lipid and carbohydrate.

epizootic: pertaining to a disease affecting many animals of a species simultaneously, analogous to an epidemic.

eukaryote: an organism in which each cell has a membrane-enclosed nucleus.

genetic code: the arrangement of nucleotides in the chromosomal polynucleotide chain (q.v.) governing the transmission of genetic information by transcription to mRNA (q.v.) and translation (q.v.) into proteins. Genetic information is encoded in the nucleic acid by the specific arrangement of four bases (two purines: adenine and guanine, and two pyrimidines: thymine and cytosine). In this language, each sequence of three adjacent bases (a codon) determines the insertion of a specific amino acid in the polypeptide. The genetic code is universal in that it appears to be the same for all forms of life studied.

genome: the complete chromosome set of an organism bearing the entire complement of genes.

genotype: the entire genetic makeup of an organism; the genes present at specific loci.

"helper" virus: a virus that aids the development of a defective virus by restoring or supplying the activity of a viral gene required for lytic infection (q.v.).

heteroduplex: an experimentally reconstituted molecule containing a segment of single-stranded DNA hydrogen-bonded to a second segment of a strand of RNA or DNA containing partially complementary base sequences.

heterologous: pertaining to different species or lack of correspondence in structure, as between an antigen and an unrelated antibody. In the latter example, either may be said to be heterologous to the other.

heterotypic: pertaining to, characteristic of, or belonging to a different type.

histone: proteins rich in basic amino acid (e.g., lysine and arginine) and deficient in tryptophan. Histones are found in combination with DNA of all eukaryotic (q.v.) cells except fish sperm. Their fundamental amino acid sequences have been conserved during evolution.

homologous: pertaining to the same species or correspondence in structure, as two chromosomes, or an antigen and its specific antibody.

homology: the fundamental similarities among structures, e.g., macromolecules in different organisms derived from a common ancestor.

hybrid: the progeny resulting from mating of genetically unlike parents.

hybridization of nucleic acid: formation of a double helix by annealing one chain of DNA to a chain of DNA or RNA by means of hydrogen-bonded complementary base pairs.

icosahedron: the geometric form composed of 20 equilateral triangular faces (the shape of some viruses).

immune electron microscopy: visualization by electron microscopy of negatively stained viral particles aggregated by specific antibody, as in the detection of enteroviruses in stool specimens. Also refers to detection of antigens in cells or tissues after reaction with antibody labeled with electrondense ferritin or peroxidase.

immunosuppression: the prevention or diminution of the immune response, as by irradiation or administration of antimetabolites, antilymphocyte serum, or specific antibody, or as in the course of certain viral infections, e.g., measles.

inclusion body: the specific acidophilic or basophilic heterogeneous virus-associated mass resulting from viral infection and visible by light microscopy in cytoplasm and/or nucleus.

internalization: the transfer or penetration of a viral particle of infectious nucleic acid across the plasma membrane to the interior of the cell. The transfer occurs by viropexis or, with some viruses, through fusion of the viral envelope with the cellular envelope.

isopycnic equilibrium gradient centrifugation: the technique of centrifugation whereby molecules or viral and other subcellular particles can be separated by equilibration at a point in a concentration gradient (e.g., of CsC1) equivalent to their buoyant density.

lectins: cell-agglutinating proteins, most of which have been isolated from plant seeds, e.g., phytohemagglutinin.

lymphokines: factors released by stimulated T cells (q.v.), e.g., macrophage inhibition factor (MIF), which blocks the mobility of phagocytic monocytes and macrophages.

lytic infection: viral infection leading to synthesis of new virus, death of the infected cell, and release of the viral particles.

messenger RNA (mRNA): the RNA molecule formed by transcription (q.v.) that is complementary to a DNA cistron and that functions as a translation (q.v.) template.

minus or plus strand: pertaining to the ability or inability of a nucleic acid sequence to function directly in translation. "Plus" strand refers to viral RNA that can itself function as messenger (e.g., picornaviral RNA); "minus" strand viruses require synthesis of a complementary (plus) strand, from which mRNA is then made, for translation to occur (e.g., paramyxoviruses, sometimes called "negative strand" viruses).

monocistronic: the expression of the genetic information of a cistron (q.v.)

as a unit (e.g., the megalopeptide synthesized in the early stage of poliovirus replication).

mutant: an organism bearing an altered gene that may or may not express itself in the phenotype (q.v.) of the organism.

mutation: a permanent transmissible change in the genetic material, usually in a single gene. The change may be in the form of a loss, gain, or exchange of one or more nucleotides in the nucleic acid comprising the gene.

naked: pertaining to a virus having no envelope.

nonreplicating subunit vaccine: a viral vaccine composed of viral particles that have been disrupted and are therefore no longer infectious (e.g., influenza virus killed by treatment with ether).

nucleocapsid: a unit of viral structure, consisting of a nucleic acid with associated protein that, in some virus groups, may be the capsid (q.v.). A virus may be a naked (q.v.) nucleocapsid of cubic symmetry, or, in either cubic or helical form, it may be enclosed in an essential envelope.

nucleoprotein: the combination of nucleic acid and attached basic proteins.

oncogene: hypothetical vertically transmitted viral genetic material that codes for the protein that is postulated to transform a normal cell into a malignant cell.

oncogenic: capable of inducing neoplasia, benign or malignant.

operon: a genetic unit consisting of adjacent cistrons that are transcribed (q.v.) and regulated together. Cistrons function coordinately.

permissive: relating to a cell type or conditions (e.g., temperature) that allow full expression of the viral genome, i.e., support lytic infection (q.v.).

phage (bacteriophage): a virus that infects bacteria.

phenotype: the observable physical, biochemical, and/or physiological properties of an organism produced by the genotype (q.v.) as it interacts with the environment.

plaque: a clear area in a monolayer of cells produced by destruction (lysis) of contiguous cells or fusion of indicator cells by several cycles of virus proliferation.

plasmid: an extrachromosomal genetic element capable of antonomous replication in the bacterial cell.

polycistron: pertaining to a giant mRNA molecule produced by adjacent cistrons in the same operon (q.v.) that specifies the amino acid sequence of two or more proteins.

polymerase: any of several enzymes that catalyze the formation of DNA or RNA from preexisting purines and pyrimidines, in the presence of DNA or RNA, which acts as a template.

polysome (polyribosome): a group of two or more ribosomes that are aggregated by attachment to an mRNA that functions in synthesis of the specified protein.

provirus: the viral gene(s) integrated into the host-cell chromosome and transmitted from one cell generation to another without causing lysis of the host cell. By the application of any of several physicochemical techniques, provirus may be rescued from its integrated state and produce infectious virus. Formerly referred to as prophage, provirus generally is also now used to describe oncogenic virus that has been integrated into the genome (q.v.) of a mammalian cell concomitantly with transformation (q.v.).

pseudovirion: a virion (q.v.) that contains fragments of cellular DNA in addition to viral DNA encapsidated by viral proteins.

recombination: formation of progeny having combinations of genes different from those of either parent. In influenza viruses, recombination occurs through the independent assortment of genetic units.

restriction endonucleases: components of the restriction-modification cellular defense system against foreign nucleic acids. These enzymes cleave unmodified (e.g., methylated) DNA from other organisms.

reverse transcriptase: RNA-dependent DNA polymerase, characteristic of RNA tumor viruses (retroviruses). RNA acts as a template (q.v.) for the enzyme in synthesis of DNA.

serotype: a subgroup within a species, e.g., microbial, determined by the kinds and combinations of constituent antigens present in the cell and detectable by antibodies of known specificity.

strandedness: referring to whether nucleic acid exists as a single or double strand of purine and pyrimidine bases, e.g., single-stranded DNA.

S (Svedberg) unit: a unit used to describe sedimentation in a centrifugal field; the unit with which particles or macromolecules settle that is proportional to their size but influenced also by density, shape, diffusion, and other physical properties.

T antigen: an early virus-specific antigen found in the nucleus of cells infected or transformed by certain oncogenic viruses.

T cell: thymus-dependent lymphocyte that mediates cellular (as opposed to humoral) immune responses and in mice possesses the surface phi (ϕ) antigen.

temperature-sensitive mutant: a conditional mutant (q.v.) that expresses one phenotype (q.v.) at certain temperatures but not at other temperatures because of inactivation of protein(s).

template: the macromolecular mold that guides the synthesis of a similar or related macromolecule.

transcription: the formation of mRNA (q.v.) by the attachment of individual nucleotides by base pairing (guanine with cytosine and adenine with uracil) and synthesis catalyzed by RNA polymerase.

transformation: the changes in the biologic function and antigenic specificity of a cell that result from integration of, and subsequent regulation by, viral genetic sequences in the cellular genome and confer on the infected cell certain properties characteristic of neoplasia.

351

translation: the formation of polypeptide chains with a specific amino acid sequence that corresponds to genetic information transcribed to the mRNA from the DNA.

virion: an intact viral particle.

viroid: a highly resistant, nonimmunogenic pathogenic agent thought to consist of a short RNA molecule in close association with cellular membrane.

virus: infectious disease-causing agent that is smaller than a bacterium, that requires host-cell biosynthetic machinery for replication, and that contains either DNA or RNA as its genetic component.

VP1, VP2, VP3: viral structural proteins, e.g., of papovaviruses.

wild-type: the reference genotype or typical form of an organism as ordinarily encountered in nature.

xenotropic: pertaining to a virus that infects cells other than those from which it was isolated, or pertaining to an antibody made in one species that is cytotropic for cells of a different species.

Index

JC virus, 19, 52, 196, 205, 251, 253,
	261
Junin virus, 31

K virus, 19
Kern Canyon bat virus, 28
Kilham rat virus, 16, 17
killer cells, 8
Kirsten sarcoma virus, 222
Kozhevnikov's epilepsy, 251
kuru, 53, 233–238, 242, 244–247, 249
Kyasanur forest disease, 136, 137

L cells, 16
lactic dehydrogenase virus, 26
Lagos bat virus, 28
lambda bacteriophage, 37
Landsteiner, K., 3, 6
Lassa fever virus, 31, 136, 137, 140, 270
latent rat virus, 16
Lentivirinae, 29, 30
leukemia, 53, 54, 221, 224, 226
leukoviruses, 28
live-virus vaccine, 287
Löffler, F., 3
Lupidon G., 157
Lupidon H., 157
lymphocytes, See B and T lympho-
	cytes
lymphocytic choriomeningitis virus,
	31, 33, 50, 51, 54, 266, 268
lymphoma, 53, 222
lysis, 5
Lyssavirus, 28

Machupo virus, 31, 138
maculopapular rash, 68
maedi virus, 29–31, 235
mammary tumor virus, 30
mapping, 37
Marboran, 316
Marburg disease virus, 28, 39, 40, 136,
	137, 270
Mason-Pfizer virus, 30, 218
mastadenovirus, 20
Matthews' classification of virus, 33, 35
measles virus, 28, 33, 51, 54, 96, 250,
	253, 286

anamnestic response, 297
antibody, 296
atypical, 295
clinical manifestation, 297
diarrhea, 97
epidemic, 296
immunity, 299
immunization, 293, 294
post-measles leukoencephalitis, 250
reimmunization, 309
subacute sclerosing panencephalitis,
	295
vaccination, 298
vaccine, 295
melanoma, 207
meningioma, 206
meningitis, 24, 120, 274, 276
meningoencephalitis, immunodefi-
	ciency, 250
messenger RNA, 16
methylisatin, 270
mink, 243
minute virus of mice, 16
Mokola virus, 28
monkey, 39, 71, 139, 182, 207, 221,
	243, 252
rotavirus, 25
mamary tumor virus, 30
monkey-adapting component of adeno-
	virus, 37
morbillivirus, 28
mosquito, 124, 141
mouse, 209, 218, 243
diarrhea, 25
hepatitis virus, 32
MS-1, MS-2, See hepatitis virus
Mt. Elgon bat virus, 28
multiple sclerosis, 50, 249, 251, 253
mumps virus, 28, 54, 120, 264, 266,
	268, 270, 286
attenuated, 300
imunization, 300, 302
vaccine, 300, 301
murine hepatoencephalitis virus, 54
myocarditis, 24, 54
myositis, 54

Nairobi sheep disease, 32
nasal swab, 273
nasopharyngeal aspiration, 273